T0341201

The Four Pillars of Portfolio Management

Organizational Agility, Strategy, Risk, and Resources

Best Practices and Advances in Program Management Series

Series Editor
Ginger Levin

RECENTLY PUBLISHED TITLES

PgMP® Exam Test Preparation: Test Questions, Practice Tests, and Simulated Exams
Ginger Levin

Managing Complex Construction Projects: A Systems Approach
John K. Briesemeister

Managing Project Competence: The Lemon and the Loop
Rolf Medina

The Human Change Management Body of Knowledge (HCMBOK®), Third Edition
Vicente Goncalves and Carla Campos

Creating a Greater Whole: A Project Manager's Guide to Becoming a Leader
Susan G. Schwartz

Project Management beyond Waterfall and Agile
Mounir Ajam

Realizing Strategy through Projects: The Executive's Guide
Carl Marnewick

PMI-PBA® Exam Practice Test and Study Guide
Brian Williamson

Earned Benefit Program Management: Aligning, Realizing, and Sustaining Strategy
Crispin Piney

The Entrepreneurial Project Manager
Chris Cook

Leading and Motivating Global Teams: Integrating Offshore Centers and the Head Office
Vimal Kumar Khanna

Project and Program Turnaround
Thomas Pavelko

The Four Pillars of Portfolio Management

Organizational Agility, Strategy, Risk, and Resources

Olivier Lazar

CRC Press is an imprint of the
Taylor & Francis Group, an **informa** business
AN AUERBACH BOOK

CRC Press
Taylor & Francis Group
6000 Broken Sound Parkway NW, Suite 300
Boca Raton, FL 33487-2742

Printed on acid-free paper

International Standard Book Number-13: 978-1-138-60132-1 (Hardback)

Visit the Taylor & Francis Web site at
http://www.taylorandfrancis.com

and the CRC Press Web site at
http://www.crcpress.com

Dedication

Jacek, you've always been beside me while writing this. Your music made my thoughts move forward and gave me the strength and persistence I needed, as it always did. I miss you.

Contents

List of Figures

Foreword

When Olivier Lazar asked me to write the Foreword for this book, I was pleased to have the opportunity to reinforce the importance of portfolio management in organizations, as a practitioner. According to my professional experience, organizational growth usually results from successful projects that typically generate new products, services, or procedures. Top managers want to obtain results and better and better business results, but many times they are not focused on the why and the how. More and more managers are aware that they need to have projects, programs, and operations to transform their strategy to execution. However, projects seem to be unlinked to the organizational strategy, and managers are unaware of the quantity and scope of projects within their organizations.

When, as a consultant, I ran a survey in several organizations about assessing their project environment, one common answer was found: "*We have too many projects.*" When analyzed, some of those projects were unnecessary. I have helped financial firms to implement portfolio management in their organizations, which was one way for them to discover that they were not investing in the right projects. Selecting projects necessary for their strategic emphasis helped to resolve such feelings. But executives need to be conscious that projects are the means to execute their strategies. The common goal is to fulfill the overall strategy of the organization. Usually, all projects draw from one resource pool, so they share some common resources.

This book explains the main pillars to be established to make executives, managers, project managers, PMO managers, students or academics understand the issues that arise in the practice of portfolio management and be able to not only practice it but also implement it in a sustainable way in their organizations.

Sound portfolio management practice can help link strategic decisions with execution. But strategic managers often lack the expertise, skills, and/or means to make their strategies concrete to deliver good results. Project managers lack the ability to understand or question strategic terms and are often not aware of expected results. Operational managers understand the need to have a project portfolio identified and under control, but they were not trained on portfolio management, so they often find it difficult to manage it correctly. All of them need to read this book carefully. In doing so, they will gain insight into how to improve their organization's performance through a dynamic decision-making process (portfolio management).

After working for organizations more than 30 years, first as a project manager, second as a PMO manager, and then as a project, program and portfolio management consultant in a wide range of industries, I have found that the portfolio management discipline is missing in many of them. The definitions, concepts, and methodology explained in this book will provide executives with the means to achieve their objectives and build and manage a portfolio that will be a true measure of an organization's intent, direction, and progress. In my experience, portfolio implementation is a management team effort. By this I mean that all managers need to believe in the practice of portfolio management to be able to dedicate the necessary effort, reflect upon, and establish an effective mechanism to make dynamic decisions. This mechanism needs to be aligned with the continuous organizational change and adapt, adopt, and apply it.

In my personal life, I need to manage a portfolio of six components—my wife, two sons, one daughter, my sister-in-law, and my dog. I needed to spend a lot of time getting to know my portfolio components better and better, including their needs and expectations (I did not ask my dog). But every year I review my family strategy, and we need to work as a team at the end of each year to visualize and plan the next year's projects, programs, operations . . . Are you using a portfolio mindset? You need to be convinced about its feasibility and buy it yourself before trying to sell it to other people. This book is written by a practitioner who always shares, asks, works, and interacts with other project portfolio practitioners.

This book is meant to represent a wide view of portfolio management practice. All the ideas and suggestions in this book are based on the author's experience, insights, and best practices. Read it carefully, apply a theory, and test it by practicing it. Revisit this book periodically and verify that Olivier's ideas work well. I am sure that he will be more than happy to receive your comments, ideas, and feedback about this book.

Remember that our careers, as professionals, never end. We are continuously learning, day by day. Perhaps this book will inspire more and more curiosity in you. Move forward and never stop reading and learning. Every day is a good day to learn, but if you read this book, tomorrow will be better for you.

Alfonso Bucero, MSc, PhD Candidate, CPS, PMP, PMI-RMP, PfMP,
PMI Fellow
Managing Partner, BUCERO PM Consulting
www.abucero.com

Acknowledgments

I would like to dedicate this book to my dearest friend and mentor Michel Thiry, whose continuous challenges and support have pushed me to go beyond my own boundaries and limitations and allowed me to get where I am and to head to where I am going.

I would like to extend special thoughts to Christophe Bredillet, who was the first to open my mind to "real" project management, and to Russel Archibald and James Snyder, who I have the great honor to count among my friends and who exemplify engagement, courage, dedication, and kindness.

A special thought as well to my friend and colleague Larry Suda, for his heart and generosity, and whose home has been one for me so many times—and where a significant portion of this book was written.

And as it's my first book (and maybe the last one . . .), I especially want to thank all of my family and friends—Jean, Céline, Elisabeth, Emilie, Alexia, Charles, Estella, Caroline, Stéphanie, Sahbi, Emmanuel, David, David and David (you'll have to guess who is who guys), Manon, Paul, Sascha, Fabio and the whole Ohana, and Jacky (who had enough faith in me to tell me one day to go beyond my perceived horizon, and I thank him for that every day)—and so many others who have supported me on this journey through all times—good, bad, and worse. I've not always been easy. And not forgetting the missed ones—Jacek and Grazyna—I hope you would have been proud.

About the Author

Olivier Lazar is an organizational architecture consultant, coach, and trainer, as well as a graduate with a master's degree and an executive MBA in strategy, project, and program management from the Lille Graduate School of Management and a graduate of the PMI Leadership Institute Master Class in 2013.

Committed to the advancement of his profession, Olivier is a managing partner and COO at The Valense Palatine Group, acting Project Management Director at Altran Switzerland, a former president of the PMI Switzerland Chapter, and he has held various global volunteering leadership roles at the Project Management Institute (PMI).

With more than 20 years of organizational governance, change, project, and program leadership experience, from both an operational and consulting perspective. Olivier has worked in a large range of industries—from corporate finance to aerospace, and from E-business to pharmaceutical and energy, where he has created organizationally agile environments fostering engagement, motivation, and performance.

Olivier accompanies organizations in their transition to higher levels of maturity through rethinking their structures and governance principles. Focusing on value and change management, Olivier helps to deliver organizational transformation.

He has been published in professional publications and has presented at a number of PM conferences around the world, including the PMI® Global

Congress EMEA and PMI® Global Congress North America. He has also been a seminar leader for PMI SeminarsWorld® since 2013.

He's also one of the very few to hold seven credentials from PMI: PMP®, PfMP®, PgMP®, PMI-PBA®, PMI-RMP®, PMI-ACP®, and PMI-SP®.

Olivier's leitmotif lies in his conviction that sharing knowledge is a major factor for global performance and common development—on both an organizational and personal level. As such, Olivier is also a member of the PMI Educational Foundation (PMIEF) Leadership Society, promoting project management as a skill for life.

Chapter 1

Introduction

It indeed took me a while to process this book, in fact it represents an entire life of applying and developing, along with my clients, colleagues, and friends, these different concepts.

As I have moved forward in my practice of organizational issues and project, program, and portfolio management related topics during my different assignments, I have noted a growing trend in the specific aspects of portfolio management, and as I have moved up in various roles—including sometimes a few executive C-Level ones—I have noticed that everything in an organization is in fact a matter of portfolio management.

Portfolio management as such represents the overall steering principles of an organization. In fact, the top portfolio manager within a company is none other than the CEO of that company (don't tell them that!).

Some other aspects also raised my interest, such as risk management. I realized very early on that if everything in an organization is portfolio management, then everything in portfolio management is about risk management. Each and every single decision we make at each and every single moment in our lives is a risk-based decision. This is how our human brain has been wired by seven million years of evolution, and this is how our human brain still operates. And that's true even more in a context of portfolio management in which what we have to deal with is the exposure of our organizations to risk—be it a strictly financial, organizational, business, or contextual risk. Whatever action we undertake, it's either to counter a threat or exploit an opportunity.

Another big trend lies within the concepts developed around agile practices. I'm personally one of the believers that agile, in terms of a project management approach, has always been around and corresponds to a proper and flexible application of the iterative project development cycles. But, more interesting than the simple application of agile project management principles, the definition of organizational agility as the development of anticipative capabilities allowing us to strengthen the sustainability of an organization by better adapting it to a fast-changing business environment, seems to be a quite interesting and engaging topic.

In the following pages, we'll dig through this very concept, but the ability to anticipate and introduce just the appropriate amount of change, when and how it is needed, is nothing other than the very definition of project management (as a global and generic framework). It's also exactly what we mean by organizational agility.

These are then the main concepts and bricks we'll assemble to obtain an integrated portfolio management framework aimed at being at once a set of communication tools throughout the organization—tools that (1) allow all layers, vertical and transversal, to speak the same language, using the same words for the same concepts; and that (2) provide a means to ensure its sustainability and build the capabilities to realize its strategic vision while remaining flexible and developing an engaging work environment to install a collaborative and performance-driven mindset and culture.

Even with all these capabilities, we are still faced with a dangerous pitfall: organizational entropy. Entropy is the tendency of any system to change its configuration from stability to chaos. And the more you complicate a system, the more and the faster its entropy will grow, and then the less you will be able to control it. In a governance system of any kind, the entropy is created by the multiplication of controls, metrics, indicators, processes, and reports that not only increase its entropy, but consequently increase also the organization's inertia. That deadly combination results in a blinding of the organization's management and decision-making bodies, triggering an illusion of mastery and a dreadful loss of productivity. Also, on the human side (which is the main asset in an organization, as we'll explore it in the following chapters), it creates a vicious cycle of demotivation, leading to disengagement and finally to a drop in performance and productivity.

But enough of introductions, let's start the journey, let's tell our story with a bit of context and background scenery . . .

Chapter 2

Context of Portfolio Management

Let's take a general look at the different layers in place within an organization. But first, allow me to define what I mean by the term "organization." It's not restricted to the concept of a company—in other words, a legally established entity. It covers any formally or informally defined cluster of co-organized, structured, and coordinated individuals working toward the achievement of a common objective, whether quantitative and/or qualitative. It can be temporary or not. A company is of course an organization, but so is a department, division, function—you name it—within that company. A project or a program group or team can also be included within that definition.

Then, what's going on in these organizations?

Everything starts with an idea, a concept: a trigger which can be a client demand, a new product or service idea, a business opportunity to exploit, or a constraint to respond to. This piece of processing is usually covered by our research and development (R&D), innovation, ideation, and other similar endeavors within the organizations. The outcomes of these endeavors result in the form of a new capability to develop, a new product or service definition, a business case for a new market opportunity, a statement for the development or improvement of an internal capability.

But this new statement, if identified as a potential value trigger, how to develop it? How to deploy it? How to exploit it? Very often, we have a very limited perspective and a short amount of information with which to answer these questions. We face here a complex problem, and we need to put in place and use a specific governance principle to reduce that level of complexity and lead us toward the realization of the expected benefits.

Reducing complexity is like "eating an elephant" . . . How do you eat an elephant? One bite at a time. Meaning, we need to cut the complex problem into smaller pieces that are easier to define, manage, and deliver; the sum of these pieces results in the delivery and creation of the new, expected capabilities. This is how programs are defined. To paraphrase Thiry (2015), programs are collections of harmonized and coordinated change actions (projects and activities) aimed at delivering benefits and capabilities not obtainable from single framed initiatives.

The components of these programs are in themselves projects which will have to be planned, executed, and delivered to create the necessary tangible means to establish the new capabilities and integrate them within the organization. Projects create the results, and programs integrate them within the organization, ensuring that the organization is able to absorb and exploit these results and *eventually* generate the expected benefits (eventually—and eventually only—because these new capabilities will effectively deliver results if, and only if, the organization is able to exploit them—in other words, to put these results into operations). Operations is the organizational layer that uses and exploits the capabilities of the company, runs the so-called "business

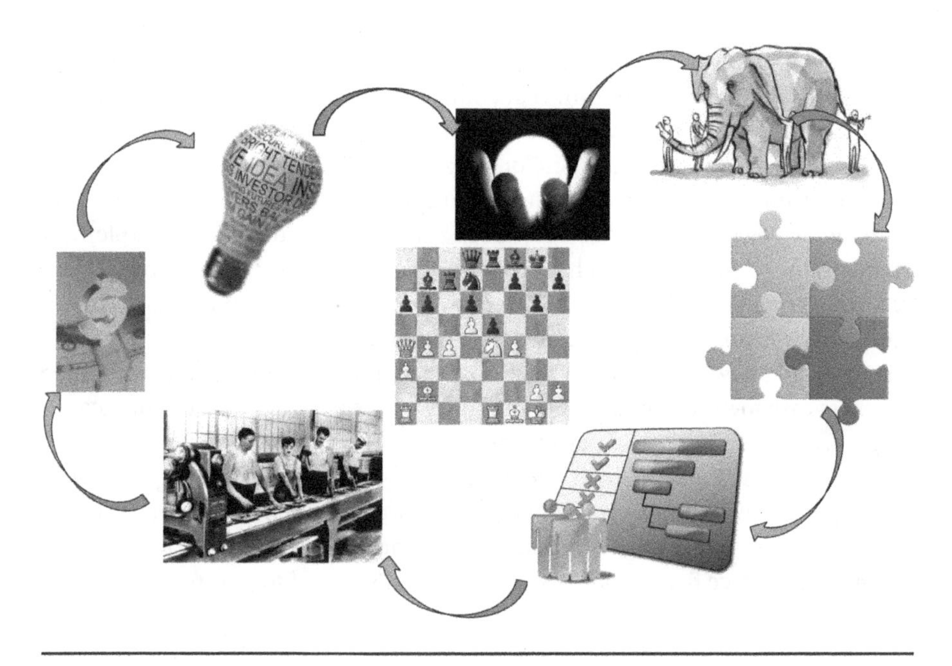

Figure 2.1 The Organization's Business and Strategic Cycle

as usual," and effectively generates from that the expected profit, income, and return on investment.

Profits ultimately are reinvested into the R&D, innovation, and ideation processes for specific projects, and the cycle can continue its iterations perpetually, like the engine in a car—but an interesting one, in that it's supposed to generate its own gas.

But keeping the image of a car, there's something missing here: a direction, an aim, a goal, and a steering wheel.

This steering wheel is essentially the strategy of the organization, which defines in which direction the R&D and innovation efforts should be oriented. It's the strategy of the organization that determines which of the outcomes of the R&D, innovation projects, and initiatives are potential benefits triggers that constitute possible means to realize that strategy, and thus are developed as programs and projects. Which of these results should be integrated within the operational level and generate the level of performance (profit, income, ROI) that will allow the cycle to continue for the next coming iterations and sustain the organization within its business environment.

This cycle is illustrated in Figure 2.1.

We can thus identify the following levels:

1. The strategic layer, steering the organization and determining the direction to aim at and the path to follow to reach that aim.
2. The operational layer, exploiting the means and resources of the organization to serve its clients and generate the necessary revenues to maintain the business sustainability of that organization.
3. The project and program layers, which build the means and capabilities to be exploited by the operational layer. Projects deliver the tangible assets and exploitable elements, and programs ensure that these elements are integrated and absorbed within the organization, securing their operability and the production of the expected benefits.

But something seems to still be missing in that picture.

We need to have in place a governance layer to support all the other layers described above and integrate them into a single framework, anchoring the realization of the strategy. We need to be able, according to the aims and goals defined within the strategy, to prioritize the allocation of resources to each of the components of the subsequent layers to optimize the usage of these resources to finally maximize the generation of performance by the overall organization.

We need to be able to decide how much of our resources to dedicate to R&D and innovation, how much of these resources to assign to programs and projects, and how much to allocate to operations, depending on our measured and assessed capacity and our performance objectives.

That's exactly what portfolio management is about. Portfolio management is the layer that embraces and integrates them all. Portfolio management is the organizational governance layer that supports the realization of the strategy of that organization by optimizing the effective and efficient allocation of means and resources to maximize the generation of performance.

The portfolio of the organization includes all the activities—present and potential—that represent a consumption of organizational resources that are intended to contribute to the overall performance of an organization. That covers operations, projects, and programs.

There is always a portfolio in any organization. The upper level of organizational portfolio is the organization itself. Of course, there may be, and most probably will be, different levels of portfolio, embedded one in each other like Russian dolls, determined by how and where the organization uses its resources and how and where it makes its profits and generates its income.

2.1 Differences between Projects, Programs, and Portfolios

A portfolio will include various elements, from ongoing repetitive daily business operations to projects, programs, and even other portfolios.

One of the key success factors for an organization in the application of its project management governance principles is to be able to differentiate the nature of these various components. In fact, the denomination "project management" for the wider practice of managing projects, programs, and portfolios is misleading. Many organizations have a tendency either to treat everything as a project or to use the wrong criteria to segregate these elements, usually by budget size or other sizable and quantitative factors.

But an incorrect identification and categorization of these organizational components can often have critical consequences. Projects, programs, and portfolios are very different animals. Managing them requires a specific set of tools and techniques and a specific set of skills and competences. If we apply the wrong tools to the wrong element, it's very unlikely we'll achieve any of the expected results.

Let's see what these differences are.

As mentioned above, projects deliver the tangible elements—assets—that will have to be exploited and operated by the organization's operational layer to obtain the expected benefits, build the desired capabilities, and generate the necessary performance, return on investment, profit, revenue, growth, etc.

Projects are then aimed to assemble tangible outcomes that are delivered to the project's sponsor by the project manager, so that the sponsor (very often a program manager) will be able to produce the expected benefits to be obtained from the operability of the project's result.

Project management standards and guides, such as PMI's *A Guide to the Project Management Body of Knowledge (PMBOK® Guide)*—6th Edition (2017a) or the PRINCE2® methodology (AXELOS, 2017), will provide you with all the insights you need to deliver these tangible results.

But the very first point to really understand here about a project is that a project as such does not produce any benefit, nor does it generate any direct performance. A project is a cost—politely speaking, it's an investment. The benefit is produced, or the performance generated, if and when—and only if and when—the organization is able to absorb the outcome of the project and exploit it, even in organizations that directly sell the results of their projects to their clients and get paid for that. The project is about creating an outcome—a tangible result. It's very different from being able to sell it or to use it.

We will assess then the success and performance of our projects according to their ability to create these tangible outcomes and results.

The transformation of a project's outcomes and results into benefits, integrating these results into the operational layer of the organization to create new capabilities and produce the expected benefits, lies within the scope of program management.

A program is a "collection of change actions (projects and operational activities) purposefully grouped together to realize benefits" (Thiry, 2004). In a program, we don't create any tangible outcomes alone, we don't manage our technical ability to deliver a clearly stated result: We have to produce benefits, we have to build organizational capabilities and manage organizational change. All these statements are mainly qualitative instead of being based solely on quantitative metrics and criteria.

To achieve these results, we have to manage the interactions, harmonization, and coordination of our various program components, rather than the technical delivery of tangible outcomes. Program management is an activity of integration and decision making more than one of execution and control.

We'll then assess the success and performance of our programs according to their ability to produce these qualitative benefits—their ability to create new capabilities within the organization.

The tools to manage things from this qualitative perspective are also very different from the ones used with projects. The *PMBOK® Guide* will not help here. A program management toolbox, such as *Managing Successful Programmes (MSP®)* (AXELOS, 2011b) or PMI's *The Standard for Program Management*—4th Edition (2017c), will be more relevant.

These projects, programs, and related activities are of course part of at least one portfolio. The aim here is to ensure that resources are properly and optimally allocated to maximize the overall organizational performance. We'll assess the success of our portfolios based on their ability to generate that performance, be it direct revenue, profit, return on investment, or even productivity; and on their ability to support the realization of the organization's strategic vision and give visibility over the evolution of the company's business environment, giving to decision makers the necessary information to navigate within that environment.

We'll assess performance and success of our portfolios based on quantitative metrics. Defining these metrics, managing this performance-oriented perspective and the effective coordination of resource, requires again a different set of tools. Here the appropriate toolbox could come from the *Management of Portfolios (MoP®)* (AXELOS, 2011a) and/or PMI's *The Standard for Portfolio Management*—4th Edition (2017b).

The nature of our components is also related to their level of complexity, or should I say to their level of complication. Analyzing these levels will contribute to determining the governance models we'll have to set up in our portfolios.

2.2 Uncertainty, Ambiguity, and Complexity

We can evaluate our portfolio components based on two factors:

1. *First, their level of uncertainty.*
   Uncertainty represents a lack of quantitative information—that is, everything related to cost, time, effort, resources, etc. The less predictable are your quantities, the less accurate are your estimates; the more risks you have identified on a certain component, the more uncertain is that component. For instance, on a building construction project, you know quite precisely what you have to build and how it will be built. You won't start any construction activity without having quite detailed plans and drawings. But you can only make a more or less accurate estimate about how much time, resources, materials, and effort you will have to invest, because these quantities depend upon risks and incidentals that will potentially appear during the execution of your project and impact it. That's precisely the example of a project with a high level of uncertainty.

2. *The second factor represents their level of ambiguity.*
   Ambiguity also represents a lack of qualitative information—that is, everything related to the very definition of the final outcome, result, or deliverable of your component, or even about the definition of the process to produce that outcome. When you need to initiate a feasibility study or a research and

development project, very often you don't know what the outcome of such a project will be, nor necessarily how to obtain that result—sometimes not even if there will, at the end, be a result. In the pharmaceutical industry, when a clinical study is undertaken within a drug development program, the result is unknown, and not even the significance of that result can be predicted. The lack of ability to clearly define the scope, the end result, and/or the executing process of a project creates a high level of ambiguity.

Combining the assessment of these two factors will allow us to determine the level of complexity of a certain component. Complexity, being *uncertainty × ambiguity,* represents the level of multiplication of unknown parameters and dimensions that must be considered and managed when handling a specific component of our portfolios, as represented in Figure 2.2 (Lazar, 2012).

When looking at the examples given above, the construction project and the feasibility study or the drug development clinical study project, we can see quite clearly that their nature is very different, and they present different if not opposite levels of ambiguity and uncertainty. The construction project is very uncertain, but certainly not ambiguous—we know exactly what we have to build. On the other hand, the clinical study and the feasibility study are very ambiguous, but absolutely not uncertain—we have already decided precisely

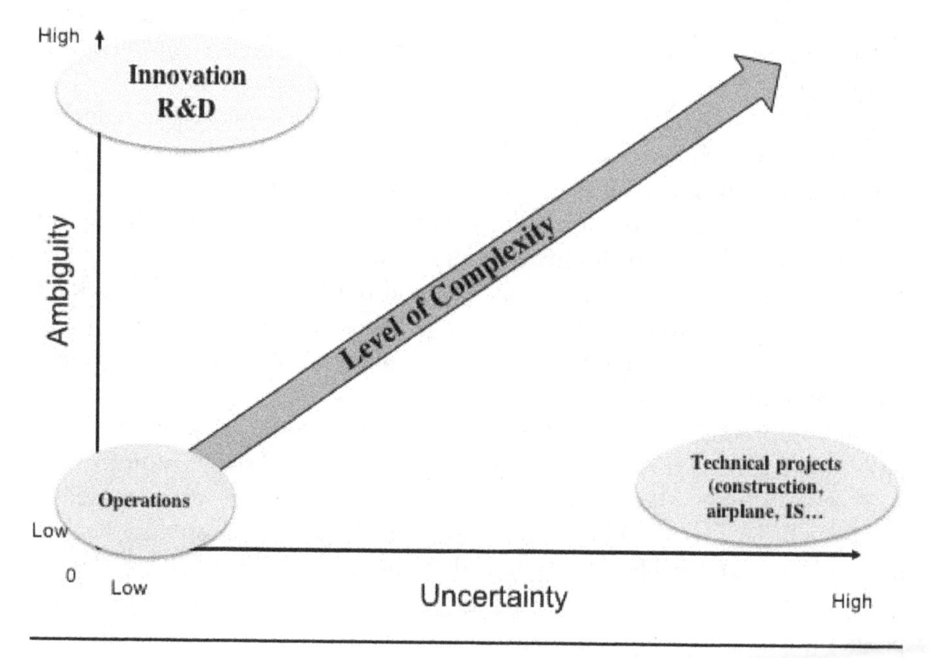

Figure 2.2 Uncertainty, Ambiguity, and Complexity

how many resources will be invested into these projects. In fact, on a component defined as a project (according to the definition given in previous sections), these two factors are mutually exclusive. On a project as such, one can't have simultaneously a high level of ambiguity along with a high level of uncertainty. A project is something that is not *complex,* but it can be complicated. The higher are the uncertainty or ambiguity parameters, the higher is the level of complication, but it can't be complex, just complicated. Sorry for the tenets of all models for "complex project management."

And it's not a matter of size. Everyone will agree that assembling an airplane is a huge endeavor. But no aircraft manufacturer will start this assembly without precise and detailed blueprints. The level of ambiguity is then very low. The engineers, technicians, and other team members know exactly what they have to do and how to do it. The only unknown part is related to the exact amount of time it will require, the amount of resources, and the risks that might disturb the planned course of the project. This lack of quantitative information creates a certain level of uncertainty.

It's the same when you build a house; you won't start digging the ground before having detailed and precise plans from the architect. No ambiguity here again, but a certain level of uncertainty.

Same again with an IT development project. You won't start developing anything before having clearly specified requirements and functionalities, be it for the whole software you develop or when applying agile methodologies aimed at splitting the development into smaller pieces (releases and sprints) whose content has to be fixed before any development can start. Agile is in fact a way to keep the inherent level of ambiguity under control, but it won't help in reducing the uncertainty. We will see that agile approaches are very similar to program management principles, but that's a different story for the moment.

When constructing our component projects, while facing a high level of uncertainty, we'll have to eliminate ambiguity by securing a clear and as stable as possible definition of the final outcome and result to create a solid rock on which we'll be able to anchor our project estimates and plans, whose aim will then be to reduce uncertainty through the application of project management processes, tools, and techniques.

Let's now consider a research and development project, a feasibility study project, or a clinical trial in a pharmaceutical company. All these examples have as a common point their inability to predict their result—or even to predict if there will be a result. This creates ambiguity. There's no possibility here to rely on a clear and stable definition of the project scope. The rock-solid foundation will then come from your ability to diminish uncertainty. If you can't predict the outcome, you must make a decision about quantities, deciding on the amount of resources, budget, and time you will dedicate to that component of your portfolio; you may then see what the end result will be, which will usually lead to making another decision which might be the trigger to launching another component within the portfolio. This kind of approach is called "deterministic estimate" in most of the reference standards and methodologies, even if it's not really an estimation process as such.

All these examples can be quite large initiatives, but as large as they are, they are just projects, not programs, implying a certain level of complication but no complexity.

On the extreme lower side, we find operations—the daily routine of the organization that presents a zero level of complexity and complication. Let's take as an example a car manufacturer. The mass production of a certain model of vehicle has no space for ambiguity, because the scope of the product to assemble is perfectly known. It's the same with uncertainty: The amount of time, resources, and cost associated to the assembly of that vehicle are perfectly known. Operations as such present a zero degree of complexity or complication.

Of course, complex components exist. There will be in your portfolio elements with a high level of ambiguity and a high level of uncertainty at the same time. But then, you're not facing a project anymore—it is now a program. Programs are complex, project are complicated. Program management is aimed at reducing and managing complexity; project management is aimed at reducing either uncertainty or ambiguity, reducing and managing complications, not complexity.

Being able to determine and differentiate the nature of our various portfolio components is essential is establishing a proper Portfolio Management Governance Model. We'll have to define which set of tools, techniques, and procedures we will apply to each component, and also which approach to adopt to pilot our portfolio in order to be able to look at the appropriate and relevant metrics and parameters supporting our portfolio monitoring and controlling activities.

If we measure our portfolio components from the wrong perspective, we will get the wrong indications. If we want to measure our project upon its ability to obtain benefits, we will not be able to see any expected result. And this is true even if our project teams are doing their best on that project. It is the same with a program: If we assess its ability to deliver a tangible outcome or generate a certain revenue or income, we have to expect to face a great deal of disappointment. Nonetheless, this program either will create a certain business capability or not.

This eventual misalignment of success metrics with the nature of the components can also lead to bad decision making, with sometimes very harmful consequences at levels beyond the very scope of the component itself. The portfolio management approach, by looking at the whole picture from the organizational and global strategic point of view, will allow us to integrate these different components of various natures and consider them within a broader perspective.

"Everybody is a genius. But if you judge a fish by its ability to climb a tree, it will live its whole life believing that it is stupid." — Albert Einstein (*Source:* https://quoteinvestigator.com/2013/04/06/fish-climb/#return-note-5880-1)

This integration and broader perspective are conditioned by the definition of a proper Portfolio Management Governance Model.

2.3 The Portfolio Management Governance Model

Portfolio management consists of:

- establishing the organizational strategy,
- assessing the capabilities of the organization in terms of available resources and investment capacity,
- analyzing the current set of initiatives and operations being run within the organization,
- making and adjusting the forecast of incoming activities, investments, and revenues, and
- reconciling these different perspectives in accordance with the strategy of the organization defined at the executive level.

It consists also of defining that prioritization model and applying it to the different portfolio components, present or future.

The Portfolio Management Governance Model is mainly divided into six sequential and iterative groups of activities (see Figure 2.3).

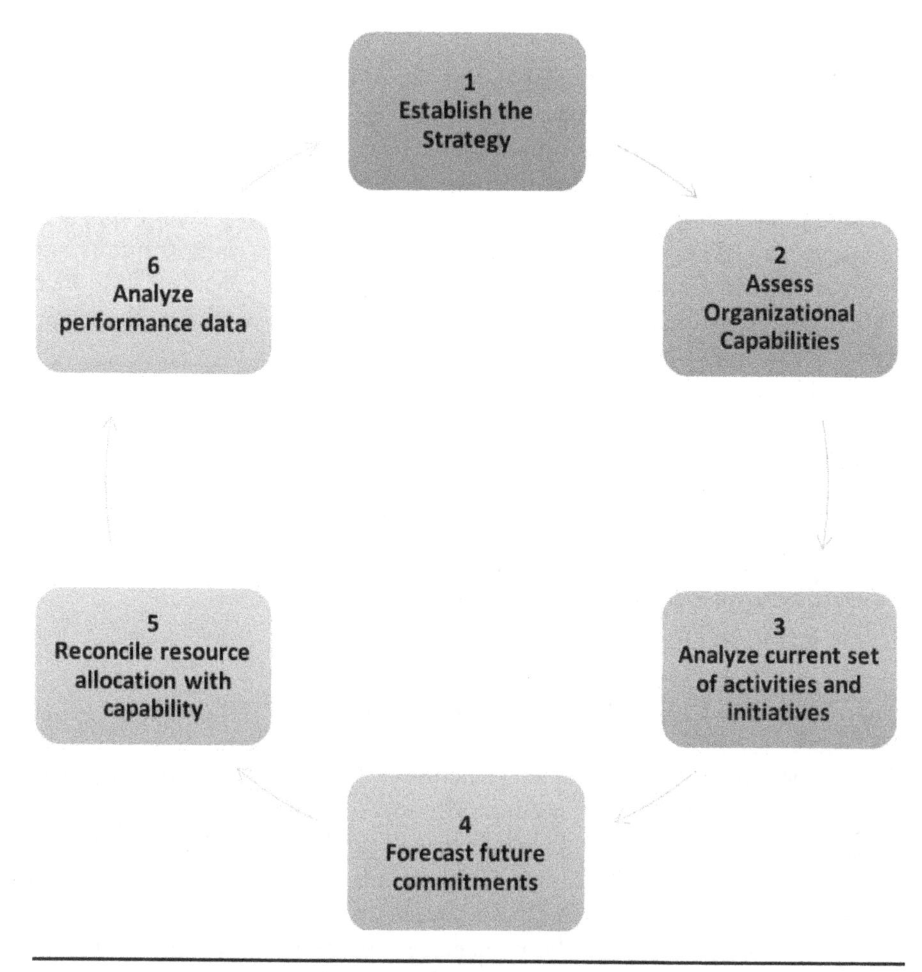

Figure 2.3 Portfolio Management Governance Model

2.3.1 Establish the Organizational Strategy

Let's be clear: It's not within the scope of portfolio management to define the strategy of the organization; that takes place at the so-called *enterprise level*, which we will not necessarily address here. But it's the role of portfolio managers to translate this strategy into a plan and execute this plan through the various actions they will put in place. Portfolio management is an activity of communication, the portfolio is what represents the strategy, and the portfolio structure and management plan are the tools used to communicate this strategy and align all components of the organization with the strategic goals.

Portfolio management will then be the channel through which the strategy will be made visible, understandable, and shared throughout the entire organization, and even outside of that organization.

Because a strategy has to be made public, everyone in the company must be aware of the strategic vision and the strategy. There's no way a strategy can be realized if it's kept in a safe whose key is only held by a designated happy and privileged few. The strategic vision is what drives people, what commits and engages them. You have to share it. Michael Porter, in this talk at the 2014 PMI PMO Symposium®, made a strong statement about this aspect: "If you hide your strategy, you'll never achieve it" (Porter, 2014).

Establishing the strategy will then be this exercise of translating the strategic vision of the organization into a tangible action plan aimed at allowing the organization to navigate through its business environment and reach its goals while securing its sustainability. Also, this strategy will require determining a certain horizon, a certain timeframe of relevancy of our actions and predictions, a strategic horizon that will define our portfolio roadmap and the delivery of the expected outcomes, benefits, and profits (see Chapter 3).

2.3.2 Assess the Organizational Capabilities

A key aspect of the portfolio management exercise is to provide the organization's decision makers with an accurate visibility over the real available capabilities. In that sense, portfolio management will be the connection between the operational and tactical layers and the human resource management layers in the organization.

We'll include in the portfolio management procedures the elements related to so-called *resource demand planning,* which will be detailed in more depth in a following chapter. But without breaking the suspense, we can say here that this resource demand planning will allow us to map precisely the current resources available within the organization by identifying exactly their kind, per nature (equipment, material, human resources) and per competences or skills (IT engineers, platform administrators, business analysts, experts, technicians, specialists, etc.).

2.3.3 Analyze Current Set of Activities and Initiatives

Analyzing the current set of activities and initiatives consists of charting the usage of these resources over the entire portfolio: which project and program, which operation is using what kind of resources, when and how. We will see in the next sections that this resource allocation and utilization can be taken from various perspectives, depending on the organizational structure in place.

2.3.4 Forecast Future Commitments

We will determine the future potential usage of these resources, based on business development forecasts and market evolution predictions.

A risk-based analysis model allocating to each potential business opportunity a weighting comparable to its expected monetary value, but applied to its resource requirement forecast and a business development lifecycle, will help to determine how much of these resources should be included in the resource demand planning exercise.

This business development life cycle (BDLC) will guide the amount of resource to be allocated to each future initiative as it moves along the cycle (see Figure 2.4).

The amount of resource allocated will be revised on a regular basis as the different business development opportunities evolve to reallocate resources from the ones failing to the ones moving forward through the cycle, until they reach the fifth stage, where they are no longer considered potential opportunities, but rather actual initiatives to be developed and executed.

The aim is to determine the gap between the current and future required capabilities to realize the strategic plan, which will come with the next step.

2.3.5 Reconcile Resource Allocation with the Organization's Capability

This is where the real and effective prioritization exercise takes place in the portfolio management practice. Often resources are scarce in the organization and we have to decide, based on the strategic vision and then from our priorities, where to put these resources.

It's a global exercise, involving all layers of the organization. This is where connections between these layers are made, this is where the interactions between the PMO(s), human resources, operations, business, and executive departments and functions occur.

As the previous steps were mainly bottom-up processes, demand driven, this reconciliation is a top-down process, capacity driven. According to the strategy of the organization and its eventual re-alignment, this step in the governance model leads to deriving, from the collected information, decisions being made and realignment of priorities, to develop the strategy for the next portfolio management cycle up to the determined strategic horizon.

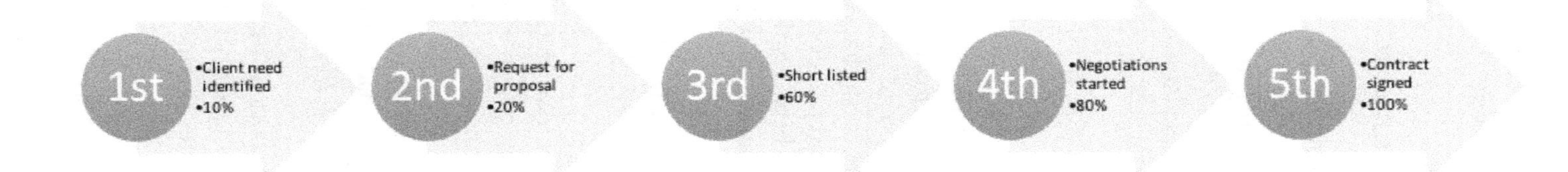

Figure 2.4 The Business Development Life Cycle (BDLC)

Time will come then to analyze the results and integrate them into the decision-making process aimed to shape the next iteration of our portfolio management exercise.

2.3.6 Analyze Performance Data

The aim of portfolio management, as we have described it in previous sections, is to generate performance. The aim of this analysis will then be to ensure that the portfolio is generating enough profit, productivity, and return on investment to maintain the organization's business sustainability and develop the necessary additional capabilities to ensure its growth.

This analysis will then be based on very quantitative indicators. We'll compare the actual business performance of our portfolio components to the targets determined in the organization's strategy. The analysis of these metrics will allow us to define to which extent we'll have to realign the strategy and/or the mix of the portfolio itself, reassigning resources, launching new components, and eventually terminating some other ones. All this with the perspective of an optimal usage of the existing capability, balancing investments with the obtained performance.

The Portfolio Management Governance Model will describe the very processes at stake for each step of its cycle, defining the various tools, methods, and techniques to be used to steer the delicate equilibrium, the fragile balance, between resources and performance.

The governance model needs to be established at the level of the whole organization; it's here to bring a guidance and consistency to the way the organization itself is governed. As such, it will be constrained and will constrain, in a kind of chicken-and-egg relationship, the overall structure of the portfolio, if not the very structure of the organization itself. Thus, the organizational context will have a tremendous impact on the definition of your governance model, shaping and being shaped by it at the same time. The chicken-and-egg paradox at its best.

2.4 The Organizational Context of Portfolio Management

The very structure of the company is an indicator of various aspects of that company: its culture, the way it does business, and even the nature of the business. Very often, transforming the business model of a company implies changing its organizational structure, or the other way around.

As we are all familiar with, there are three different families of organizational structures (and everything in between, or their combinations).

The first one is the so-called classical *functional structure.* It's a vertical, siloed chart. It's determined by the different activities performed within the organization, each of these activities having its own territory, called *function, branch, department, division, line,* you name it.

Denominations vary widely from one company to the other; the naming you will be using doesn't really matter, in fact.

Each of these silos is dedicated and defined by the area of business or activity it covers—for example, a human resources department, a marketing and sales division, or a whatever technical business line. People working in these departments are experts in their domain, and it's what they do 9 to 5, five days a week, and nothing else. In a functional organization, Marketing doesn't do any HR, HR doesn't do any IT, and IT doesn't do whatever the other departments or divisions are doing. Each of these functional lines are composed of experts of that specific domain and are headed by a single clearly identified and designated individual, very often an expert in that domain as well. This person holds full hierarchical authority over the resources of that department. The communication and authority lines are very much vertical from the top to the bottom levels, and the delivery and reporting is then bottom-up only, as described in Figure 2.5.

This particular kind of organizational structure can be seen as very constraining and inflexible, developing a lot of blockers in the information flow and in the decision-making process. Indeed, as we'll explore more in depth in the following chapters, a functional organization, by its nature, generates a considerable level of organizational inertia, and it might seem not really fit for the installment of a project management framework, especially if we are targeting a reasonable level of maturity integrating, within this project management framework and culture, domains of program and portfolio management, triggers and consumers of fast-paced decision-making processes and organizational agility.

Some of the major pitfalls in a functional structure will be related to the impermeable barrier between the branches. It blocks transversal communication and generates difficulties when it comes to integrate various components of a project produced by different branches. A famous example comes from an aircraft manufacturer having to completely rewire a large part of an aircraft prototype because the wiring teams in the different assembly sites did not communicate with each other. The plugs aimed to connect the systems of the different parts simply did not match one with each other.

Also, there's no clear difference between "project work" and "business as usual." In fact, when you're working in a functional organization, you don't see which part of your work is exploited by your managerial level as part of a

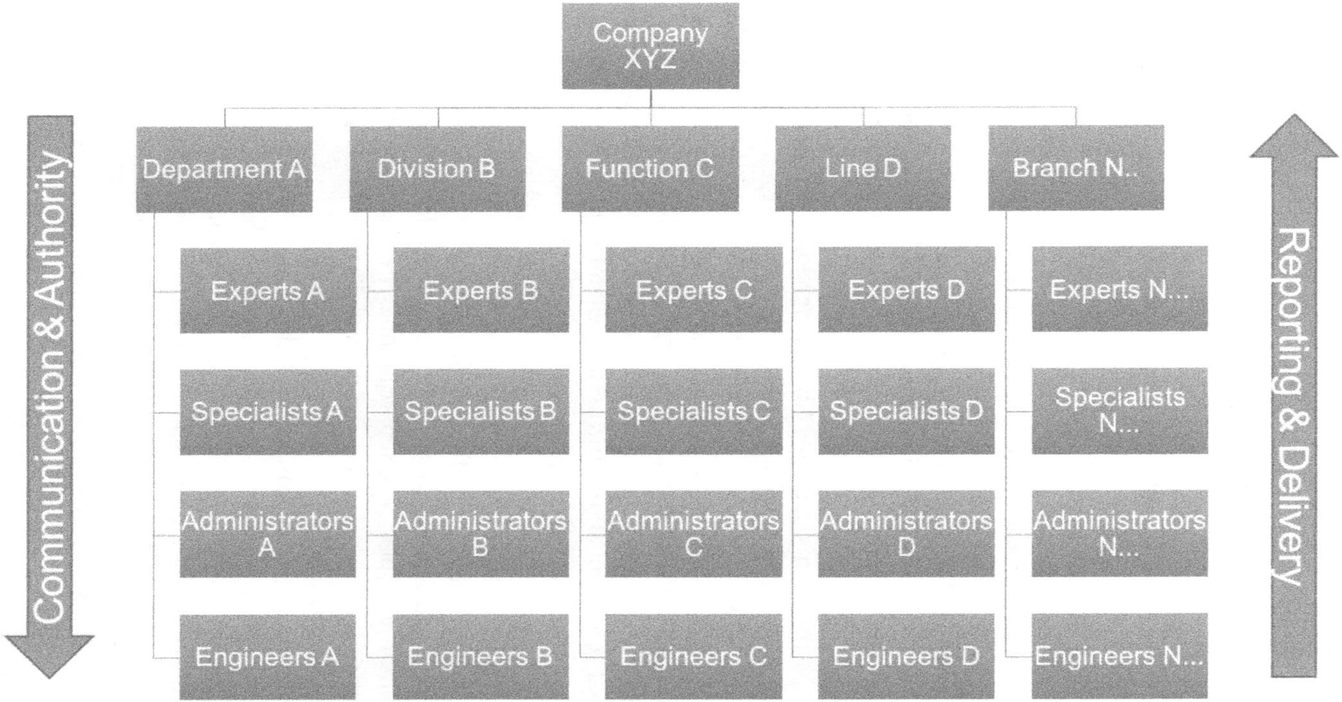

Figure 2.5 The Functional Organizational Structure

project, and you're not even necessarily aware of your contribution to any project or who else in the company is working for (not "on"; the distinction between "on" and "for" here is intentional and quite important) that project. Your best friend could be working in the neighboring department, you could be having lunch together every day and working for the same project without even knowing it.

Another difficulty resides in the total lack of control and authority over the project resources. In fact, there are no project resources as such, only people producing outcomes that need to be extracted and integrated somehow in a patchwork which we will call a project. That difficulty will be dealt with by the manager designated as the "project coordinator," sometimes also called "project facilitator."

But the picture is not so bad from every point of view.

From the perspective of the resource person, when you're hired within that kind of organization, you're hired for your potential and not only for your current competences. And it's in the interest of this organization to develop that potential.

An IT department in a company hiring a developer will have an interest in having this developer learn about new technologies, languages, and systems. The more the organization develops their employees' potential, the more value the employees represent for the organization, and the more value the organization represents for the employees. Also, such employees will evolve in a stable and somewhat secured environment, allowing them to develop a straight career path.

But indeed, a functional structure is not ideal for implementing and exploiting a project management framework, culture, and mindset.

However, there are usually some good reasons for choosing to shape an organization in a functional way. First of all is the size of the company. The bigger it is, the more it will tend to be functionally organized, because it's simpler to manage, monitor, and consolidate. Most of our portfolio reporting is based on consolidation principles. The reporting lines are clear, and the hierarchical relationships, as well as the related and derived decision-making processes, are obvious and unambiguous.

Another rationale for choosing this kind of structure will be based on the nature of activities and type of business conducted by this organization. A functional organization's business is mainly conducted through its operations. It's a company making its revenues by selling products issued by its on-going daily activities, such as a car manufacturer, a postal service, a government service, or a grocery shop. It doesn't mean that these types of business can't be structured differently, but they will represent the most common examples.

Because portfolio management is about controlling and managing the quantitative performance of an organization, it's very important to determine where that performance (in fact, the money) comes from and how the resources are spent. In a functional organization, money comes from the operations, directly selling products and services they produce to customers who pay for them. Operations being also the major resource consumers, the portfolio structure will then match the organizational structure, and the portfolio governance model will often follow also the organizational chart of the company. The highest level of portfolio management (and often the only one) will be assumed by the top executive level—the CEO.

On the other hand of the organizational continuum, we find the so-called *projectized organization.* The chart of the projectized organization is also silo based, constituted by vertical branches clearly separated one from the other. But this is where the similarity with the functional organization ends. The silos and branches of the projectized organization correspond in fact to individual projects or programs, put under the full hierarchical responsibility and accountability of a clearly identified project or program manager who reports directly to the executive level of that organization; the people belonging to these branches are all the different resources, persons holding all the necessary set of skills and competences the project or program manager will need to achieve the expected results, making it a mix of various people and different resources (see Figure 2.6).

The communication and authority are, as in the functional structure, a top-down flow, and the reporting and delivery goes from the bottom to the upper levels. The clarity and simplicity of reporting lines are also similar. But, as opposed to the functional structure, it's supposed to be a heaven for project managers. No integration problems anymore, everyone works full-time for the project and the project only, all participants are united in the same team, and the project information flow is obviously boxed within the project cell. And, the icing on the cake, the project manager has full control and authority over the resources allocated by the portfolio manager to the project.

Ideal situation then? Actually, not for everyone . . .

When you are one of the resource persons there, you are not hired for your potential. You are hired for what you know and can do right now. Developing your potential is actually not so much the concern of your manager. If you have personal development expectations, you'd better count on yourself for that. In this context, your value for the organization is your knowledge and your competence, so you won't be so keen to share it with others. Lessons learned, knowledge sharing, and capitalizing on experience are difficult in a projectized organization. Another pitfall is hidden in the very definition of a project or a program: They are both temporary endeavors (PMI, 2017a, 2017c), which means that the branch, in fact the project or program team, is also a temporary thing. When the project or program is over, the team is dismantled, and people are reassigned (or not) to other portfolio components, and not necessarily together. The team-building exercise is then perpetually in reconstruction. These teams are hard to be pushed in the Performing stage of the Tuckman Cycle (Tuckman, 1965).

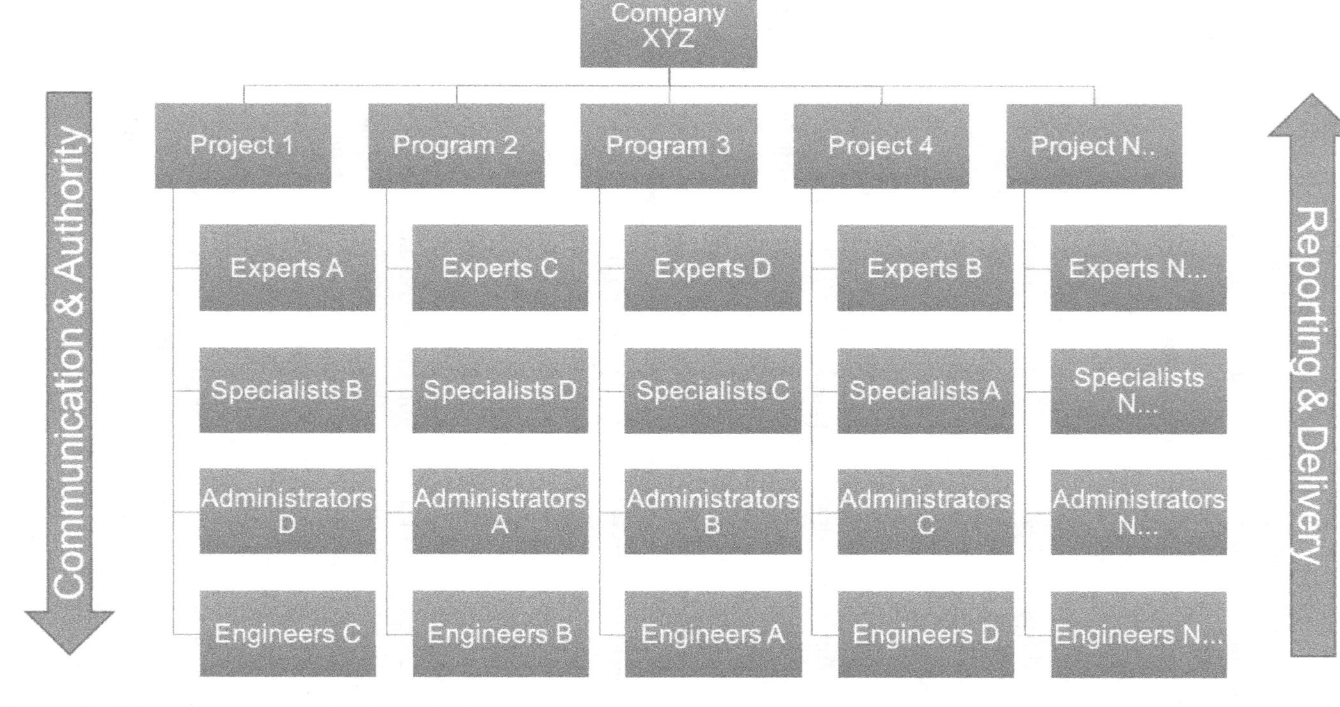

Figure 2.6 The Projectized Organizational Structure

Despite its ups and downs, like the other structures, choosing a projectized organizational chart is no coincidence. It's defined by the nature of the business conducted by the company. We see projectized structures in companies directly selling the outcomes of their projects to their clients. Projects are their daily business. Everything they do is client oriented and everything is a project—that's what makes the income, consumes resources, and defines the strategy, if ever there's a need for a strategy other than simply "respond to customers' demands."

Most of the time, small companies, start-ups, and service-based organizations will be structured this way. But on occasion, we can see larger and more complex organizations considering themselves as "projectized." I remember a client, a military organization which most could consider as a functional structure: Everything there was a project, from acquiring equipment or painting a building to deploying forces on an operation field. And for each activity, there was a project manager appointed. The only variable was the rank of the person designated as the project manager; a private was project manager for the painting job, and a senior officer was project manager for the field operations. It implies that often in a projectized organization, being a project or program manager is just a role and not a title or a position.

Again, there are good and bad sides when it comes to work in a project management framework.

Then, the solution should be somewhere else, somewhere in between, maybe within the *matrix organizational structure.*

The matrix organization is a functional organization. The structure with specific domains and the vertical authority and communication channels remains. But there is one major difference that changes the whole paradigm: The boundaries between the silos have been broken, which means that transversal communication is now not only allowed but encouraged (see Figure 2.7).

This means also that the project- or program-related communication can go through the different departments involved within the actual initiatives. There's then definitely some good sides project-wise: People know what they have to do and why they have to do it, which part of their work is part of the project and which is still within the scope of the company's business-as-usual, objectives are clear, and as the team members are connected within a formal project or program team, there are no integration problems anymore (at least, fewer of them). Speaking of people, they benefit from all the advantages of the functional structure in regards to their development, retention, motivation, and personal evolution.

So, is that organizational heaven? Not quite . . . Here also, we find some downsides; first, the complexity of the organization is very much increased. When one person works on a project, she or he reports hierarchically to two

Figure 2.7 The Matrix Organizational Structure

managers, the functional manager (often referred to as the *resource owner*) and the project or program manager. When that person works on three projects, she or he reports potentially to four people, and so on. Then the number of conflicts related to resource allocation, availability, and engagements rises drastically and creates a very high level of stress on the stakeholders, and of course it makes the exercise of resource allocation and prioritization to be challenging, making the portfolio management framework even more essential and critical to the efficiency and effectiveness of the organization.

How then to deal with these negative consequences of what seems to be a fertile ground for the implementation of a performant project management framework?

The key is hidden in the variation of the spreading of authority between project/program and the function. It will allow us to determine if we will implement either a weak or a strong matrix.

In the weak matrix, the function (or resource ownership) bears more authority and decision-making power than the project or program. A company using a weak matrix usually makes most of its business through its operations, having to develop projects and programs to create the products and services provided to clients by the operational layer. The projects serve the operations. Most of the resources are consumed by projects, but the income is generated by the operations. We'll find here somehow the same kind of companies as the examples given while addressing the functional organizational structure, but maybe with a higher level of maturity or a different culture.

In the strong matrix, the projects and programs are directly at the service of clients, generating the company's income and revenue. The functions are there to support the projects and programs by providing resources—competent, developed, and motivated. The functions serve the projects and programs, which are at the core of the business.

There is a third version of the matrix structure, the so-called *balanced matrix*, in which the authority is equally spread between the function and the projects. I think there's no need to explain why it's the worse choice ever. When no one bears the authority, there's no authority at all, and the decision making creates potentially unlimited conflicts and organizational inertia. This structure is to be avoided at any cost.

In terms of portfolio management, the matrix organizational structure is bound by a lot of parameters and imposes some constraints for reporting and monitoring.

In a matrix structure, be it weak or strong, we will have to provide a double entry to our portfolio. The project layer will need to have an overview from its perspective, structured around projects as unit components, seeing how much of resources and what kind of resources are consumed by each of them. The functional layer will need to see which projects and programs consume their resources, having as consolidation nodes their pools of resources. Among these two perspectives, the major entry which will be used to drive the organization and run the decision-making process in the portfolio management exercise will depend on the variation of the matrix, which is driven by the way the company conducts its business.

2.5 Various Perspectives of Portfolio Management

When speaking of portfolio management, and in particular the differences in its applications compared to programs and projects, we have to detail the differences of perception and thus the differences of mindsets to be considered and adopted while embracing the portfolio level.

2.5.1 Time and Performance

As I have already mentioned—and as widely defined in the various related dictionaries and standards—projects and programs are temporary endeavors. Often even projects and programs place their life cycle in a totally different temporality than that of the organization in which they are conducted.

Projects or programs don't respect or fit within a fiscal, organizational, or yearly calendar. They have their own specific calendar, which can span two or more organizational temporalities. When a financial or accounting controller asks a project manager how many resources and what budget the project will consume before the end of the year, and before the end of the following year, and to provide estimations and forecasts fitting between January 1st and December 31st, that project manager usually has to make a lot of complex calculations and decompositions of the project's resource allocation to be able to provide a response.

Let's be clear from the beginning: The temporality of portfolio management is yearly, the same as the organizational financial and fiscal temporality.

In fact, portfolio management is the tool used to reconcile this organizational temporality, balancing the one of the operational level with the one of projects and programs, and vice versa.

By collecting the various data from the different components and mapping these data onto the roadmap, portfolio management will allow the organization to develop a sort of translator between the enterprise level and the portfolio components, allowing us to construct the necessary visibility and predictability to execute the organization's strategy. By the same channels and means, portfolio management will make the strategy understandable by those who will have to deploy it throughout the various programs and projects and thus facilitate the establishment and absorption of new organizational capabilities and their exploitation by the organization's operations to generate the expected level of performance.

2.5.2 Decision Making

The decision-making perspective is also very different from the one leading projects and programs, even if there are also major differences in decision making between projects and programs.

The aim of projects is to deliver a tangible outcome, ready to be used by the project sponsor, program manager, and organization's leaders, so the decision-making process would be mainly oriented toward, and driven by, the necessity of either strictly abiding by the defined requirements if the level of ambiguity is low, or the necessity to maintain the project within the established quantitative parameters if the level of ambiguity is high.

The questions to ask to ourselves here are: "Am I delivering within the 'Project Management Bermuda Triangle' (time, scope, quality)?" "Am I delivering a result that will be a benefit trigger for my sponsor, meaning will my sponsor be able to integrate that result within the organization's daily business?" "Is my result compliant with my quality standards and performance metrics?"

In a program, the most important aspect is the creation of organizational benefits or establishment of new organizational capabilities. These capabilities are the necessary means to achieve the organization's strategy. The decision-making process will then be driven by these qualitative considerations, value driven, aimed at obtaining these benefits through the optimization of the mix of program components, all being related by their direct contribution to the expected benefits.

In that sense, program management, as a discipline, is very close to agile principles. In agile, the focus is put on delivering the highest possible level of value by prioritizing the requirements to be covered, aiming at quality first before quantitative considerations.

Also, in a program, the decision making is mainly about optimizing the connections between components rather than managing the components themselves. What matters here is more harmonization and value rather than efficiency and performance.

The questions we ask ourselves here are: "Am I producing the expected benefits?" "How can I exploit opportunities and maximize the level of benefits to be obtained from my program?" "Are all my program components contributing to obtain these benefits?" "Are the capabilities delivered to my organization aligned with its strategy, and will these capabilities be able to generate the expected level of performance?" "Is my program roadmap fitting within my organization's strategic horizon and producing benefits at an acceptable pace for my stakeholders?" "Have I properly planned and executed the transition of my projects' outcomes within the operational layer of the organization?"

In a portfolio, the focus is on performance: maximizing performance, optimizing usage of resources, predicting and anticipating the future needs and engagements.

The decisions here are aimed at finding the right balance between the usage of resources and assets, the creation of these assets, and the level of performance to be achieved to sustain the organization in its business environment long enough to enter the next strategic planning cycle.

Then it's quite a different focus and perspective than in the other layers of projects and programs. The questions we ask ourselves here are: "What is my ROI (meaning balance between the resources I use and the outcomes I create)?" "Do I have the right kind and the appropriate amount of resources?" "Are my outcomes (projects' results, programs' capabilities, and operations' constant delivery) aligned with my organization's strategy?" "Am I not wasting resources on initiatives and activities that are not directly contributing to sustain my organization?" "Are all of my components delivering their results within the time frame of my organization's strategic horizon?"

Of course, the governance model you will establish in your portfolios (see Section 2.3) will have to describe the overlaps and connections between these different layers in terms of decision making. Even if the perspectives and perceptions are quite different, there must be some overlaps and common concerns which will help to interconnect the components within a common strategy and establish a vertical alignment among these components. These overlaps are very often determined by quantitative considerations, such as the budget perimeter of each component. Each component manager must be the owner of the decision-making process within her or his allocated budget, escalating to the upper level everything that goes beyond that perimeter in terms of consequences or strategic focus. That strategic focus will also help in creating these decision-making overlaps, this time with a qualitative perspective, allowing us to ensure that at the end, we have created an integrated strategic framework, aimed at one goal: realizing the organization's strategic vision and ensuring its sustainability within its business environment.

2.5.3 Delivery and Strategy

Because projects and programs are creating the necessary means and the utilization channels of these means to realize the strategy, a portfolio represents a strategy in itself. The strategy defined at the organizational level has obviously to be translated into actions. As Henry Ford said, "A vision without execution is just hallucination."* The very development of that strategy, as we will explore further in Chapter 4, is mostly based on the identification of risks and the definition of the risk management strategy to address them, countering business threats and mainly exploiting business opportunities. Then even if, as we have already discussed, programs and projects are delivering different outcomes (tangible results for projects and business or organizational benefits or capabilities for programs), what counts at the very end is to make sure that the strategy of the organization supports the realization of its vision, and all value triggering opportunities have been exploited, maximizing that value creation and securing the organization's sustainability.

2.5.4 Portfolio Management Agility

As we will explore further in the following pages, and especially in Chapter 6, properly conducted portfolio management is a factor of organizational agility.

Sure, but what does that exactly mean? What does portfolio management have to do with agile principles?

Often, when speaking of agile, we think of methods, tools, and techniques to manage software development projects. But agile is much more of a global mindset to adopt than a set of rules and guidelines. It's a perception and a perspective oriented toward the creation of value, consideration of the major importance of establishing a collaborative relationship with the stakeholders who will have to bear the burden of the changes our projects and programs are about to create, and the overall idea that everything we do should contribute to create value, which is expressed not only in terms of performance and compliance metrics, but also in terms of enhancing the overall business environment.

Let's first give a look at the fundamental values of agile, as expressed in the Agile Manifesto (Collective, 2001):

- Individuals and interactions over processes and tools
- Working software over comprehensive documentation
- Customer collaboration over contract negotiation
- Responding to change over following a plan

The underlying statement here explains that we have to put ourselves into a mindset oriented toward the creation of value and being open to change,

* https://www.goodreads.com/quotes/155966-vision-without-execution-is-just-hallucination

managing with flexibility. Once applied to the context of portfolio management, we could interpret these values as follows:

- "Individuals and interactions over processes and tools," and "Working software over comprehensive documentation" (even if one doesn't develop any software, agile is not only for IT) means that we have to put the interests of our stakeholders, instead of the governance models and the managerial indicators, at the top of our strategic priorities. What you do and why you do it becomes more important that how you do it. Which of course doesn't necessarily mean you have to leave apart the various compliance and conformity constrains you might encounter. It simply means that delivering outcomes that do not trigger any value creation for our stakeholders is a waste of resources, as efficient and compliant that waste could have been.
- "Customer collaboration over contract negotiation" describes the need to establish a strong and permanent link with the operational part of the organization, which will have to exploit the results and outcomes created by projects and capabilities developed by programs in order to generate the expected level of performance allowing the organization to feed itself and sustain its own existence (see Figure 2.1).
- "Responding to change over following a plan." This one might seem totally in antinomy with the very principles of portfolio management. Isn't portfolio management about creating a plan for how to run the organization up to its strategic horizon and beyond? In fact, Yes . . . and no. Yes, in the sense that portfolio management is about defining a target to reach within the strategic horizon, and plan (in fact anticipate) a course of action (a strategy) which will lead us there. And no, because defining a route to reach a destination doesn't mean to keep moving forward in a straight line whatever occurs during the journey. We will have to adapt to the evolution of our business environment and adjust the defined route, which doesn't mean the final destination has to be changed. Many routes lead to Rome.

In practice, exerting portfolio management is about making these constant adjustments to the strategic route, avoiding icebergs (business threats), and exploiting the currents that could accelerate our course and allow us to reach our destination in better conditions (business opportunities). Change is the soul of portfolio management, as it is of program management. These two disciplines have in fact a lot in common with the application of agile principles in the sense of the necessary flexibility it induces and the corresponding mindset which needs to be adopted.

Chapter 3

The First Pillar: Organizational Agility

3.1 What Is Organizational Agility?

Portfolio management is an agile endeavor, requiring adaptability and openness to change to achieve one of its major justifications: the ability of the organization to be adaptable enough to survive the constant evolutions of its business environment and potentially create an environment surrounding it in which that organization would occupy a leadership position, ensuring development, growth, and sustainability. These benefits can only be achieved by obtaining a certain level of organizational agility, meaning introducing that agile type of mindset and functioning within the very governance structure, processes systems, and even culture of that organization, gaining precisely that flexibility and adaptability.

Organizational agility is the ability of the organization to pursue its strategic vision and realize it while anticipating the evolution of its business environment and adapt its strategic roadmap and related governance to this evolution. In other words, organizational agility will be the ability of the organization to introduce change when and how it's necessary to secure the achievement of its objectives and its business sustainability. Organizational agility, more than a pillar of portfolio management, is the outcome of its proper application.

3.2 Organizational Inertia

Achieving organizational agility, even with the best and most properly applied portfolio management practices and principles, often encounters slowing, if not blocking, obstacles and challenges.

One of the major pitfalls or obstacles is in fact the result of a very natural phenomenon: everything takes time.

Everything takes time because we live and evolve within a physical world. Even information transfer uses physical media to circulate, which materializes the inherent abstract substance of information and slows it down.

That natural phenomenon is what we used to call *inertia.*

The basic definition of inertia, as given in *Merriam-Webster* (n.d.), is stated as follows: "A property of matter by which it remains at rest or in uniform motion in the same straight line unless acted upon by some external force." Meaning basically that any physical entity subject to a force resists that force. This resistance is a factor of its mass and the level of that force. OK, but what does that have to do with business and organizations?

Inertia is also applicable to organizations. It certainly won't surprise anyone if I say that within an organization, everything takes time. Why? Simply because an organization is a physical entity, in which information, decisions, actions, and feedback go through physical media, including the people themselves.

We can define *organizational inertia* as being the lag in the implementation of strategic decisions. We can even evaluate it quite accurately by measuring the amount of time it takes for a strategic decision made at the executive level of the organization to cascade down throughout the different layers of that organization, starting to be operationally implemented, and the feedback of that implementation to go back up to the executive level.

3.3 Factors of Inertia

Organizational inertia is created by a set of various factors, the first one of them being human. Or should I say, the first factor is humans themselves. But, hopefully, we can't get rid of humans, and that's certainly not something we would wish for; even if the level of automation created by the evolution of technologies and the development of more and more intelligent systems raises the number and the complexity of the tasks being automated, the human factor is the essential factor of an organization.

3.3.1 An Organization: Such a Thing Does NOT Exist

Let's come back to the roots for a moment. What is really "an organization"? What does that mean?

We have already seen a definition of an organization earlier in the book. An organization is a "formally or informally defined cluster of co-organized, structured, and coordinated individuals working toward the achievement of a common objective, quantitative and/or qualitative." What is the most important word in that definition? "Individuals," meaning, "people." Yes, people.

Everything that is done in an organization is done by people, for people, with people. We often hear statements such as: "The organization has decided . . . ," "The organization does this or that, doesn't do this or that . . . ," "The organization works this or that way and not that or this way. . . ." Actually, an organization doesn't do anything. An organization doesn't define or decide anything. No. There is no sentient and self-sufficient entity hidden somewhere at the upper floor of an office building, making decisions and guiding our lives toward an obscure objective and a pre-definite fate. That thing does not exist. An organization as such is a concept, it's a mental model elaborated to justify and gather the consolidation of the efforts of a group of people in creating something designated to respond to a need expressed by other people. An organization—such a thing does not exist in any tangible reality, and each time we refer to an "organization" in these pages, we'll in fact refer to that group of people.

That being said, what's wrong with people? Nothing, besides the fact that people (human beings) have their comfort zones. These comfort zones are situations and mindsets in which they have found a certain stability and continuity. The problem with stability and continuity is that in our dynamic environments, they condemn organizations (these groups of people) to stagnation and to the obsolescence of their products and services, and even of their business models. That obsolescence comes faster and faster, putting in jeopardy the ability of the group (the organization) to create wealth to sustain its members (people), and thus their ability to become consumers of their own goods and services, creating a vicious cycle of decreasing demand and supply, leading at the end to the dismantlement of the group. Then organizations need change. And people prefer the comfort of stability and continuity; they are naturally resistant to change. And there's nothing wrong about resistance to change—it's a normal human trait. We are all resistant to change, we all have the need for a certain degree of stability. It means that although we all have a certain ability to absorb a given level of change in our environment, we have a given tolerance to disturbance.

3.3.2 Resistance and Limited Ability to Change: The Sponge Effect

Let's imagine a sponge. Place this sponge into a bowl of water, squeeze it (meaning constrain it), and it will then absorb a given volume of water from the bowl—a given volume and not a single additional drop of that water. That's because the sponge has a limited ability to absorb a given volume of water.

It's exactly the same with people and change in their environment—they can absorb only so much of transformation of their environment. And when this limit is reached, people resist, making the change impossible to be absorbed at the level of the group, of the organization.

As indicated in the previous paragraph, there's nothing wrong with that, it's a factor we have to integrate into any endeavor undertaken within an organization.

That level of tolerance to disturbance will introduce a lag in decision making and in the execution of these decisions, because to ensure that a change will be integrated, we have to wait for the people to be ready and available to absorb the change to come. Resistance is created when one wants to go beyond that tolerance to disturbance. Like in a car, if you want to turn the steering wheel too much and too fast, you will feel that your movement is stopped by a physical limit, put here on purpose to block a drastic change in direction. Your car will resist, and if you want to go beyond that limit, it's more than certain that you will lose control, damage it, or both.

We then have to respect these limitations, which of course slows down the implementation of the strategic decisions being made in the organization, contributing directly to the increase of organizational inertia.

3.3.3 Organizational Entropy

An additional factor also comes into the picture when we talk about the factors of organizational inertia: entropy.

What is entropy? It's also a natural phenomenon, and it's also initially a physical one. It's the tendency of any system to change its configuration from stability to chaos.

That's basically why we get old and die. Aging is the consequence of the entropy of our body's cells.

Entropy means it's impossible to maintain the stability of a system (any system) for a long period of time—long being, of course, totally subjective. And the more complex the system will be, the more difficult it will be to preserve its stability. Entropy exponentially increases as we add parameters into the system.

It's what happens when your computer becomes slower and slower as time passes, obliging you to reset the system from time to time (at least for non-fruity computer users), because the more software, browsing histories, and cookies you add, the more parameters have to be integrated to stabilize the system, and the more entropy it creates within your computer.

Also, when making an estimate of time or effort for a given project activity, we use a parametric estimation model, with which we can see that the more parameters we add to the estimation model, in fact the less accurate it becomes.

Entropy results in a loss of control, a loss of visibility and in the means to anticipate or even to react to the evolution of our business environment by losing accuracy in the perception we have of that environment, internal or external.

3.3.4 What Creates Entropy and the Illusion of Mastery?

Entropy of course impacts organizations. An organization is mainly defined through the interactions of the people within the group, making it a system, and quite a complex one.

These interactions behave like the parameters in our estimation models: The more interactions you add, the more you need to measure and track these interactions, and the more you introduce parameters into your organizational system. In fact, the more we increase the complexity of the organization, the more controls we put into its governance system, the less we can see what is happening in that system, and the less we can control it. But a very common bias in many companies consists in adding even more controls, indicators, and metrics. The more we lose track of the organization, the more we'll have a tendency to tighten the controls over it, thus increasing the overall entropy. In many organizations, we can see a multiplication of reports, metrics, and key performance indicators (KPIs) aimed precisely at regaining control over the events and their consequences, but in fact contributing to increasing the entropy of the organization, contributing to a loss of mastery. Trying to control and measure each and every process within the organization creates a false sense of control and an illusion of mastery.

I once ran a workshop in an organization in which people had to fill in a timesheet describing everything they were doing, with a granularity of 15 minutes, every day. I asked, "How long does that take?" The answer was "about an hour." Every day. There were at that time 2,500 employees subject to that control. This means that 2,500 times one hour, times an average of 220 days per year, the calculation came to a rounded total full-time equivalent (FTE) of 270! Knowing that one FTE is usually equal to 1.2 headcounts, it gives us the amazing number of a workforce equivalent to 324 people, dedicated only to filling in a timesheet that, of course, nobody was really reading . . . 324 over 2,500!

Another example comes from an organization for which I had to give a precise opinion on their portfolio management governance system. Their structure was made of three layers. Their portfolio monitoring dashboard included 147 indicators, times three (one for each level of the organization, and of course not the same indicators at each level). I was very embarrassed when they asked me, "What more can we do?"

I can think also of another department in a company that had an indicator to measure what they called "conformity of compliance."

All of these additional controls and metrics, aimed to measure how we do things more than what we effectively produce, not only introduce complexity and thus entropy into the governance systems, but also—by losing visibility over the real value triggering work and the outcomes of that work—dilutes the purpose of that work.

Joseph Juran, noted management expert, said, "Without a standard, there is no logical basis for making a decision or taking action. Most companies understand this to some degree, but many persist in measuring performance by the wrong standard—using unsubstantiated or ineffective metrics that ultimately lead nowhere" (Juran, 1964).

That dilution of the meaning into non–value-adding activities not only creates a counterproductive waste of scarce resources but also contributes to demotivating and disengaging people in the organizations. Disengaged employees are less productive and more resistant to change, contributing even more to increasing the overall level of inertia in the organization. Employee engagement increases productivity by 21 percent and profitability by 22 percent (Gallup, 2013).

3.4 How to Overcome Organizational Inertia?

3.4.1 Simplicity and Parsimony

Then what can we do? Do we have to endure organizational inertia and deal with it? No.

Inertia can be reduced; even if you will never be able to eliminate it completely, you can try to optimize your organization.

The first optimization triggers will reside in your governance processes. The key element here is simplicity, and even parsimony.

Simplicity first. Define simple and straight-forward processes, with a streamlined flow of information. Don't include too many loops within your processes. Also keep in mind the inherent difference between data and information. Data is the raw material of information; information is a set of data that are actionable, leading to a tangible and, if possible, immediate reaction from the person who receives it. If it's not leading to action, then it's not information, it's noise.

And more than anything, ensure that your process leads to a tangible result. Something you can touch, measure, and clearly identify as a distinct element. Because the most important fact is to be able to measure the outcome of your action, more than how these actions were conducted.

Of course, in a lot of industries, compliance and conformity are important. Regulatory constraints are very strong in banking, pharmaceuticals, energy, etc., and they need to be assessed, monitored, and validated. But there is no point in producing a result that will not be a value trigger for the organization and its stakeholders, even if that result is obtained in a very compliant way. Compliance and conformity are indeed important, but making sure the organization is producing the right result for the right purpose—this is what motivates people, what gives sense to their effort and encourages them to uphold their engagements.

Parsimony now. The famous French writer Antoine de Saint-Exupéry said once, in substance, "Perfection is achieved, not when there is nothing more to add, but when there is nothing left to take away" (Antoine de Saint-Exupéry, n.d.).

Use only the indicators and processes you need, and question the processes and indicators first before questioning people if the outcomes are not satisfactory.

For the indicators, the basic questions to ask are, "Where does the information come from?" "Where does the information go and to whom?" "Is it really information or only a set of data?" In other words, "Is it actionable?" If the answer to only one of these questions is, "I don't know," or "I'm not sure," or if you have to pass the data to many organizational layers before finding any relevance into it, then most probably you can remove that indicator. In the example that I mentioned above of the company having put in place a 3 × 147-indicator dashboard for their portfolio management and monitoring, to the question they asked me, "What more we can do?" I replied, "We won't do more, we will do less." And after few months of efforts, we ended up with a unique dashboard for the whole organization, instead of three different ones, with 29 indicators only, measuring what was meaningful only and avoiding spending time on measuring *how* things were done rather than *what* was really done. Again, quality assurance is important, no doubt, but it's useless if it applies to something that is not at firsthand a value trigger.

These simplicity and parsimony principles apply as equally to indicators and metrics as they apply to processes.

Encyclopedia-like guidelines, methods, and frameworks are as useless as they are time consuming. Not only will people not use them if they don't perceive their value or if their utilization takes more time and effort than they produce results, but if there is a strong compliance requirement in the organization, they will develop ways to go around, leading to an additional effort and the loss of possibly meaningful and precious information. If a process, a guideline, or a tool is not used or is even misused, don't question the people, question that process, tool, guideline first. And if something is not used, but the outcomes are still achieved, disconnect it. If nothing happens, don't put it back.

3.4.2 Optimizing the Organizational Structure

Another aspect to look at to optimize the organization and reduce the level of inertia sits within the organizational structure itself.

Obviously, a functional organization will present a higher level of inertia than a projectized one.

The vertical-only communication and delivery channels create bottlenecks for the circulation of information and thus add delays and filters within the decision-making, execution, and monitoring flows. By multiplying the levels of filters, it can also create a distortion within that information flow which can sometimes be perceived as a kind of "black hole" in the center of the organizational chart, eating the top-down strategic communication and eating the same way the bottom-up feedback for the execution levels.

The typical situation which I encounter in this kind of functional organization is described by the executive level in such terms: "We define a strategy, based on partial information, which is very difficult to collect; then we cascade down that strategy, but we can't see the results of its execution, and it's already time to rethink a strategy because of the dynamic of our business environment."

And in the same organization, the operational level says: "We escalate a lot of data on what is really happening on the field, but we don't see the effect of our feedback on the re-alignment of the strategy." Because of the disconnect triggered by the various perceptions and the distortion induced in the information flow, these organizations are in danger of being simply stopped in motion when their level of inertia overcomes the evolution rate of their business environment.

It's what I call the *Titanic Syndrome.* The crew has seen the iceberg coming, but due to the level of inertia of the Titanic, they have not been able to avoid the catastrophe. And as the image of the Titanic illustrates well, there's no such thing as "too big to fail."

The projectized organizational structure is not necessarily better, as it is usually applicable to a limited number of business models, which are mainly oriented toward full external customer service, and thus have a rather limited strategy to develop.

As we will explore more in depth in later pages, their portfolio structure is also rather simple, often a unique basket with all activities aimed to this single perspective, and their portfolio monitoring will mainly be based on resource allocation and short-term profits, with the risk of losing a long-term perspective and of measuring resource utilization rather than overall value creation.

Projectized organizations are often at the opposite side of the parsimony/simplicity continuum compared to functional structures, but they encounter difficulties due to the excess of parsimony and simplicity in their usually simplistic governance models.

Although exceptions exist, I have personally rarely encountered projectized or functional organizations with the right balance of measurements and processes compared to the level of optimal control which would be necessary to a proper and efficient portfolio management exercise.

One alternative to this situation, with its pitfalls, could be to establish a matrix organization. The matrix kind of organization, be it weak or strong, will indeed support the optimization of communication channels, contributing to reducing the organizational inertia by flattening the organizational charts and shortening the decision-making flows. Just remember that weak matrices keep a certain verticality in the decision-making process, especially about everything related to resource allocation, which in the context of portfolio management can be counterproductive; and instead of reducing that inertia, it can maintain it at the level of a functional structure.

Matrix organizations also present a specific risk, as these types of structures usually create quite complex organizations with various dimensions, be it in terms of reporting lines, communication, decision-making, and consolidation keys for the portfolio management exercise.

Of course, as already mentioned, the organizational structure has to not only be aligned with the nature of the business model in place within the organization but also fit with the culture—some cultures allow people to feel more at their ease in certain structures than in others. A cultural misalignment could generate resistance to the implementation of executive decisions. Structures misaligned with the business model would lead to higher levels of entropy, caused by an unnecessary level of governance complication or, conversely, a governance model not strong enough to allow an accurate monitoring of the business initiatives and environment. It would also not allow the development of the appropriate portfolio structure and consolidation nodes.

Anyway, whatever would be the chosen organizational structure, the main factors influencing the direction taken to use this optimization trigger should be conditioned by essential factors such as:

- organizational culture
- business model
- appropriate portfolio structure
- relevant portfolio governance model

But more than anything, the most efficient key in reducing organizational inertia lies in the ability of the organization to develop models and governance structures in favor of embedding a culture of collaboration within that very organization.

3.4.3 Developing Cooperation

What does "embedding a culture of collaboration" mean, and why do I insist on that aspect being one of the most important factors for the reduction of inertia?

Because, as we have already discussed several times on the previous pages, an organization is an integrated system. One part cannot work in isolation from the others; what happens at one end of the organizational chart has an impact on the other end. And what is true for the organization as a whole is also true within each of its governance layers:

- Within a project, each element of the work breakdown structure is tightly related to the others. Each work package has to deliver a distinct outcome or result, which needs to be integrated with the others within the framework defined by the scope of the project and the scope of the product, solution, or service to be delivered.
- It's even more sensitive within a program, in which all components are susceptible to share some resources, but all these components are mainly supposed to contribute directly to the overall goal of the program and creating the expected capability, obtaining the defined organizational or business benefit, and also contributing to, and at least taking into account, the integration of that capability into the business as usual of the organization. In other words, a program's components are supposed to be change triggers within the organization—a change realized through the program itself. Programs are nothing more than change enablers. This interdependence and imbrication creates a profound need for a collaborative environment, the sole success factor able to assure the absorption of the change by the concerned stakeholders.
- At the portfolio level, collaboration and cooperation are also essential, as success depends deeply on the optimal spreading of resources throughout the components of the portfolios and eventually among the different portfolios in place within the organization.

In fact, collaboration is also a trend which very often is totally unnatural.

A lot of companies create compensation and reward systems for their employees based on individual performance, because it's supposed to be motivating and challenging for the employees and foster performance. Actually, what happens is exactly the opposite. In an integrated framework—and we find these frameworks in any kind of organizational structure, from functional to matrix to projectized, at different levels—what really matters is the final result, the overall objective, far more than the individual distinct sub-products.

Managerial models based on individual performance create a set of separated clusters, put one against the other, but with clear and often impermeable limits. If the performance is measured and rewarded at an individual level, then the limits of that performance have to be clearly defined and even enforced, creating small boxes of bounded accountability, with as many interfaces as you have boxes supposed to interact one with each other, creating in fact as many interfaces between the boxes. And guess what is important between two of these boxes? The interface, of course! If a deliverable within a project can't move from one work package to the other, or if a project result cannot be transitioned to the organization's operational level to be integrated within the business as usual, then obviously the value will not be realized even if the individual result or outcome is perfectly assembled. Then if the interfaces between the elements or components are so important, what to do? Obviously, develop a process to handle these interfaces. But with an interface process, we need corresponding metrics, indicators, and guidelines, and of course let's not forget to include an interface process manager who will be held accountable for the (in)effectiveness and (in)efficiency of that interface process . . . and you will need as many of those as interfaces you have created within the organization.

But if we multiply interfaces with related processes, systems, metrics, guidelines, and accountable persons, what is the concrete impact on the organization?

First of all, we create as many boxes of bounded accountability, which not only don't help in solving the interface problems, but multiply them. If you had an interface problem between two elements, A and B, and you add a supposed-to-be "connector," which we will call C, in between them, you create, in fact two, interface problems: between A and C and between C and B. and most of the time, it increases your costs, because you have to pay for C.

Then it also has as a consequence the multiplication of processes, metrics, and guidelines to increase the entropy of the organization. The more we add parameters into a system, the faster it will tend to return to chaos.

Finally, an individualistic managerial model will inevitably lead to increasing the inertia of the organization, triggering again the Titanic Syndrome.

The solution to that issue might seem quite simple, but very often it is culturally challenging: enforcing collaboration by reversing the performance-based reward paradigm, encourage cooperation, and promote collectivism instead of isolated competition, which leads inevitably to an increase in the level of entropy, inertia, and paralysis of the organization, and at the very end to is disappearance.

3.4.4 Breaking the Traditional Model of Uniqueness of Accountability and Encouraging Collaboration

In a very first step, as opposed to what is preached and taught by many managerial text books, and leadership and management courses, it is imperative to break the rule on the uniqueness of accountability in the organization, at every level, and even between the different management and governance levels.

We are all familiar with the conventional Responsibility Assignments Matrix, based on the RACI principle. There are many others, but the RACI is the most common of all.

Let's go through a quick reminder about the basic principles:

- "R" stands for "Responsible," the doers. Responsible people are the ones who actually perform the tasks listed in the matrix. You can designate as many "R" persons as you need for a particular task, activity, or component.
- "C" stands for "Consulted." If you are designated as being consulted, it means that the doers—the responsible—are expecting a kind of input from you that will be necessary to complete the work. You can put as many "C" persons as you need for a particular task, activity or component.
- "I" stands for "Informed." If you are designated as informed, it means that you are expecting an input from the responsible party of a particular activity. You can put as many "I" persons as you need for a particular task, activity, or component.
- Now, last but not least, the "A." "A" stands for "Accountable." To simplify, accountable are the ones who get punished if the "R"s don't do their job.

In French, in the translation of the RACI model, "A" means "In Authority" ("*en Autorité*"), which in my opinion, represents quite well what "Accountable" means in that context. And the common usage of the RACI model tells us that there can be only one, unique entity accountable for a particular task, activity, or component. And in addition to that, we are told that you cannot be at the same time accountable and responsible, because the accountable hold a certain level of authority over the responsible, who report to the accountable (see Figure 3.1).

This is where problems usually start.

If we want to dig a bit deeper into that, we need to clarify what Responsible and Accountable mean.

As said, Responsible people are the doers, but beyond just executing a set of activities, these people need to embrace a certain level of engagement within this execution. What would be the purpose of executing anything if we don't feel concerned by the quality, effectiveness, and efficiency of our results and the way we obtain them? Wouldn't the opposite be a clear demonstration of personal disengagement and demotivation? Indeed.

We could then define Responsibility as:

- Responsibility is accepting ownership of the specific activities within the defined roles and taking the initiative to deliver the agreed goals, objectives, and requirements of that activity.

	Activity A	Activity B	Activity C	Activity D	Activity E	Activity F	Activity G	Activity H
Stakeholder 1	R	C	C	I	I	R	I	R
Stakeholder 2	A	A	A	R	R	A	R	A
Stakeholder 3	R	R	I	A	A	I	A	I
Stakeholder 4	C	C	R	R	I	I	R	C
Stakeholder 5	I	R	C	R	I	C	C	R
Stakeholder 6	R	I	I	I	R	I	I	I
Stakeholder 7	C	C	I	C	C	I	R	I
Stakeholder 8	I	I	C	I	I	R	I	I

Figure 3.1 Classic Responsibility Assignment Matrix Using the RACI Model

And Accountability as follows:

- Accountability is the ability to make empowered decisions within a defined perimeter of authority and assume the consequences of these decisions.

We then define an obvious overlap between these two components of commitment: When you are designated as responsible for a certain activity, you also bear a certain accountability for the outcome of that activity.

By developing principles of shared accountability, related not only to the decision-making power conditioned by the control of resources and budget, but also to the execution of the concerned activities and outcomes to be delivered, we commit, and from this commitment comes engagement, motivation, involvement, and a sense of responsibility in the most noble meaning—having a feeling of the consequences of your actions, decisions, and their broader implication on the overall organizational system.

Expanding the accountability to a common ground helps to reduce the number of unnecessary layers of processes and metrics that increase organizational entropy and inertia, increasing the ability people have to react to incoming changes in their business environment, being then able to oversee the impact of their actions and decisions on the value creation process and not only on the delivery of a single limited piece. For example, by following the standard RACI model, when developing an information system, one might say that the transitioning process—consisting of training the future users of that system and eventually coaching them—is not the business of the project manager; that integrating that capability into the organization and making sure that everyone has been trained and uses the system properly over time—once this system will have been deployed—is the accountability of the project sponsor, or program manager. In absolute terms, that might be right. The role of a project *manager* is essentially to deliver the result of the project, respecting the classical Bermuda Triangle of project management—on time, on scope, on specification, and on budget; while the role of the project *sponsor,* who's very often a program manager, is to produce the benefits expected to be obtained by the organization from the projects' results. But if that transition of the projects' results within the operational layer of the organization has not been foreseen while defining the scope of the development project—taking into account the needs of the affected stakeholders and the sources of their factors of resistance—there is little chance that the program manager in charge of obtaining those benefits will ever be able to successfully implement the new system. Especially so, in that often, the manager of the development project is not held accountable for the success or failure of that implementation. Then why would that project manager introduce some aspects that he or she would not control and that could impact

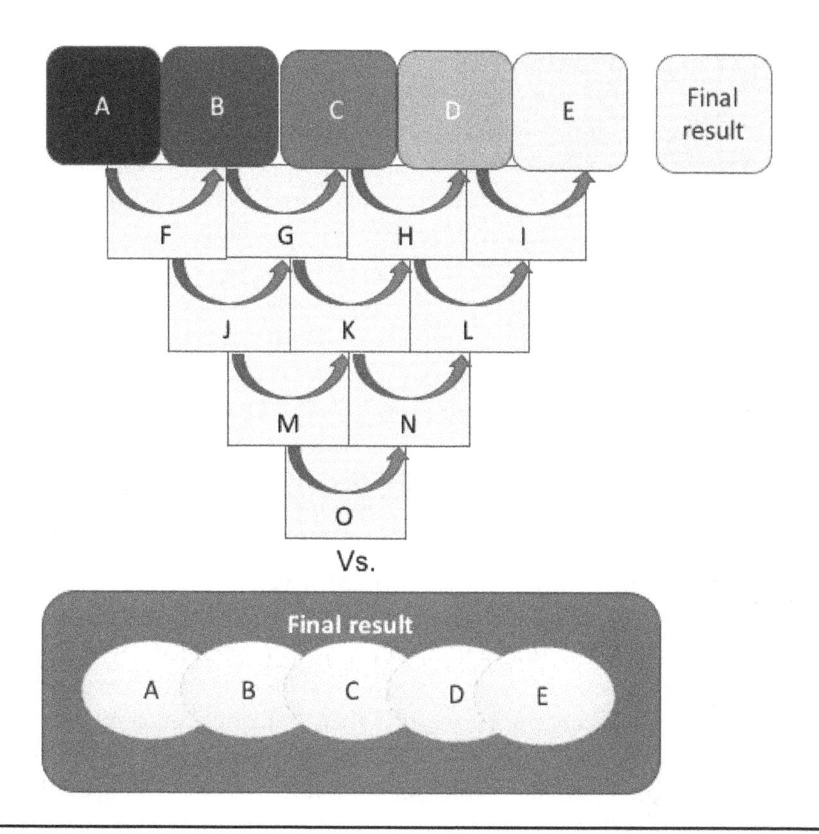

Figure 3.2 Bounded vs. Shared Accountability, Entropy and Inertia vs. Simplification and Agility

the project's scope? Anyone being rewarded on the basis of their project delivery would certainly not be doing that.

Because integrating these concerns transforms threats into parameters, allowing development of a framework aimed at success, instead of boxes defined to find the flaws when failure comes.

It also supports inducing within the organization a mindset of collaboration, mutual support, and trust, triggering a virtuous leadership cycle leading to performance and efficiency (see Figure 3.2).

3.5 Triggering a Collaborative Mindset Within the Organization

A collaborative mindset can be triggered by setting up a few simple strategies.

3.5.1 Establish an Appropriate Incentive System

When individual incentive is constructed mainly around an essential part of individual performance-based reward, on the top of which is added a marginal layer of collective results—70 percent individual and 30 percent collective as an example—it then makes sense to reverse that model and put the emphasis on collective results rather than individual ones. Even if it's still important to keep a reasonable level of individual incentive to challenge and encourage each one to go the extra mile on their own, it's important that the energy people dedicate to their performance benefit the group as a whole and not solely the unique individual. As in my example, if you make the incentive model 70 percent collective and 30 percent individual, you will make collaboration rewarding and make it in the interest of your employees to cooperate.

3.5.2 Expand the Field of Accountability

While breaking the RACI rule of uniqueness of accountability, you also expand the consequences of one's decisions beyond the field of their sole responsibility. As a result, people will have to ensure that their individual outcomes are properly defined and their colleagues are effectively able to receive that outcome. Expanding the field of accountability makes it safe to dedicate your efforts and resources to your colleague's performance, as it serves your interests as well to foster their success and doesn't damage your individual performance as it would in a bounded accountability model.

Construct a RACI matrix, but allow multiple "A"s in the matrix, allow "R"s to also be "A"s, apply a proper definition of Responsibility and Accountability (which will ensure that people who are designated as being responsible are also accountable for their work), and add a column for the end or overall result where everyone will be "A," as described in Figure 3.3 (see also the Lexicon at the end of this book).

Share it with others, and also apply a principle of reciprocity, which should be documented in such a responsibility assignment matrix. This reciprocity principle should be as illustrated in Figure 3.4.

This reciprocity principle states that if working on two contiguous elements, Pierre is Accountable and/or Responsible for element A, and Mary for element B, then Mary should be designated as being Informed on element's A RACI matrix, and Pierre as being Consulted on element's B RACI matrix. By doing so, we enforce the cooperation between the two teams and make sure that they are linked by a formal connection of mutual accountability and an obligation of communication.

	Activity A	Activity B	Activity C	Activity D	Activity E	Activity F	Activity G	Activity H	Overall Result
Stakeholder 1	R/A	C	C	I	I	R/A	I	R/A	A/R
Stakeholder 2	A	A	A	R/A	R/A	A	R/A	A	A/R
Stakeholder 3	R/A	R/A	I	A	A	I	A	I	A/R
Stakeholder 4	C	C	R/A	R/A	I	I	R/A	C	A/R
Stakeholder 5	I	R/A	C	R/A	I	C	C	R/A	A/R
Stakeholder 6	R/A	I	I	I	R/A	I	I	I	A/R
Stakeholder 7	C	C	I	C	C	I	R/A	I	A/R
Stakeholder 8	I	I	C	I	I	R/A	I	I	A/R

Figure 3.3 RACI Matrix with Shared Accountabilities

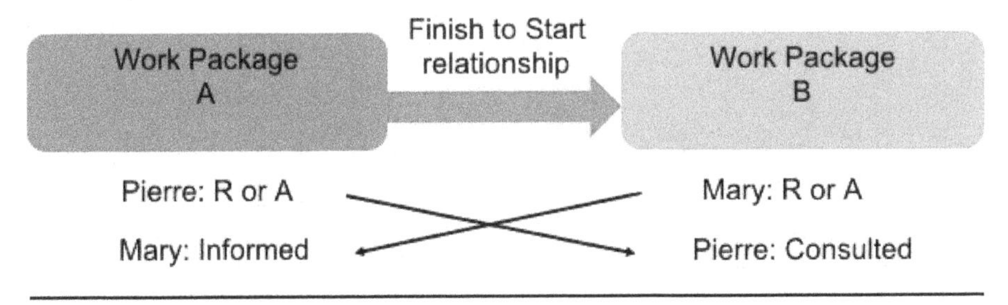

Figure 3.4 Reciprocity Principle between Two Components and/or Activities

Formalizing this helps to overcome the counterintuitive nature of cooperation and collaboration.

The concrete application of these principle does not consist, of course, in burning the RACI matrices, but rather in expanding them by breaking the two basic rules of the RACI matrix:

- When considering the definitions of Responsibility and Accountability defined above, it implies that when one is responsible for a certain activity, one should also be accountable for this activity. That sounds obvious, but in many organizations I have worked with, responsibility did not go along with accountability, which resulted in situations such as delivering a system without guarantying that the system will work—"I'm in charge of developing it, not testing it"—which indeed leads to clusters and a culture of blame aimed at covering oneself and making sure not to be punished when (not if) the project fails. "R" and "A" definitely should come together and be inherent one to the other.
- There also must be a column to represent the sum of the activities, the whole initiative, project or program, in which everyone would bear an "A" for accountable, formally representing the shared accountability among team members toward the end result, and toward the creation of value expected from that result.

3.5.3 Reward Collaboration, Make It Worthy, and Make Individualism Unprofitable

What exactly does it take to collaborate in an organization?

Collaboration is about supporting the effort of others, supporting their performance and contributing to their success, so their success becomes yours and your success becomes theirs.

Resources are limited, constraints very often lack flexibility, so if you want to support others with limited resources, it means giving up some of your own resources to support the effort of your colleagues, which inevitably will come at the expense of your own performance. In other words, would you be keen to slow down your own project to help the project of your neighboring colleague who encounters some issues? In an individualistic working environment, certainly not; damaging your own performance would mean taking money from your pocket to put it in your colleague's pocket.

It is then very important to make it safe to collaborate, and to do so, the entire performance measurement system has to take into account both the global context of the organization and the overall optimization of resource usage instead of the individual achievement. As such, it is important to put the organization into a value management framework, one aimed at measuring the value created and assessing performance and success based on value created for the organization's stakeholders rather than on absolute quantitative performance. Of course, quantitative measures are important, but quantitative performance is useless without qualitative achievement and value creation. Then there must be a reward for collaboration, even if that collaboration comes at the expense of an individual performance. Encourage openness of information sharing. Do not punish your employees because the performance of their projects deteriorates, but rather encourage them to raise issues, reward anticipation instead of reaction, and reward those who support others, especially if it costs them on their own projects and activities.

I must admit that this style of collaborative management is easier to put in place in a matrix organizational structure than in a functional one, and it is almost impossible to promote in a projectized one which measures the worth of the individual on his or hers capacities, skills, and competences, limiting their interest in sharing those assets, and creating a naturally competitive environment.

3.5.4 Communicate and Spread Your Strategic Objectives and Your Strategic Vision

The core insight in a collaborative mindset is to make people understand the global picture, sharing a common vision and, whenever possible, highlight their contribution not only to their individual perimeters of specific projects, programs, and activities, but to the realization of the organization's strategy.

Michael Porter, in his talk at the PMI PMO Symposium® in 2014 (Porter, 2014), highlighted the fact that if you want your strategy to be successful, you need to share it, you need to make it public and use it as a tool to leverage engagement and motivation. One of the channels to reach that aim can be the establishment of a project management office (PMO).

3.6 The Role of the Project Management Office in Portfolio Management and in Organizational Agility

So what about the PMO in all of that?

It comes as a double-sided blade. On one hand, the PMO can be a factor of agility for the organization, but on the other hand, if not properly structured, used, or positioned, it might very quickly increase the overall level of entropy and inertia and be more of a burden than a factor of organizational maturity.

To be a real positive force in the portfolio management effort, the PMO must be a tool of simplification and collaboration—its role becoming one of gathering data and consolidating information, ensuring the stability of the various governance processes, making sure that the connection between projects, programs, and portfolio is established and maintained. By enforcing and stabilizing these connections, the PMO can be the channel through which the executive level of the organization feeds its strategic decision-making process and communicates these decisions throughout the organization.

But, if the PMO acts like the guardian angel of processes and guidelines, behaving like the organization's police station rather than its control tower, then it introduces an overburden of complication and complexity, contributing to the overall entropy of the system and increasing inertia. That kind of situation occurs when the PMO is used solely as a quality assurance tool, overlooking the application of project management guidelines and methods and owning these processes while securing them within a framework of performance indicators, mostly related to measure how things are done instead of what is really done.

I don't say here that an organization shouldn't perform any quality assurance on its project management processes, I just say that it shouldn't be the only virtue of a PMO, or even that this quality assurance role shouldn't necessarily be assumed by the PMO itself.

Often, in our world of "project-based" organizations, the establishment of a PMO is seen as the Holy Grail of organizational maturity. If you have one, it means that you are a grown-up organization, that you know how to manage your projects, and you can be proud of it. In fact, it's a bit more complicated than that.

In order to be a value creation factor for the organization establishing it, the PMO must be constructed on a rock-solid business case, either responding to a current and particular need or filling in an identified gap generating an unacceptable level of opportunity cost—and eventually, the icing on the cake, contributing to reduce the level of inertia and entropy.

There are as many definitions of a PMO as organizations that have one or even think that they don't have one, because there's always some sort of PMO within an organization. There's always someone, somewhere assuming that role, keeping the *PMBOK® Guide* inside a desk drawer and having developed this or that template for a project charter or a program benefit register. I often say that the very first form of PMO in an organization is the cafeteria; it's the place where everyone gathers, talks, and exchanges information about what is going on within the organization; and that the first skill set of a PMO manager is the one of a bartender—listening, seeing the big picture, and intuitively knowing what the customer needs to drink.

There are also many functions that could be assumed by the PMO. From directly managing projects and programs, to providing a methodological support and owning the project management information systems, to delivering trainings and coaching to the projects and programs teams, or as mentioned above, to controlling the status of projects and programs (both from a qualitative and quantitative perspective) and providing an overview of dashboards, metrics, and indicators to feed a decision-making process and provide a level of control (or isn't that an illusion of mastery?).

Actually, the main obstacle in the role of the PMO in exerting effective portfolio management is embodied within the very denomination of the project management office as such. The word *project* itself limits the action of the PMO to within the scope of the kind of initiatives described as projects and programs. It's the same reason for which I haven't entitled this book *Project Portfolio Management,* because it's not only about projects and programs. Portfolio management is about everything that happens within the organization, everything that consumes resources, generates revenue and profits, and of course contributes to realizing the strategy of that organization.

As the portfolio embraces the daily business or operations, and portfolio management encompasses the distribution, allocation of resources, and prioritization of all organizational activities to ensure the creation of a performant and sustainable integrated framework, then the PMO also has to evolve and embrace these operational levels and have access to them, not only as an information input, but as a supporting and monitoring body.

The project management office has to evolve to a higher level of maturity, one of an enterprise management office (EMO). Does that mean the PMO and portfolio managers have to remove the word "project" from their vocabulary? No, indeed not. But they have to enlarge their perspective and the scope of their actions, responsibilities, and accountabilities.

Because portfolio management implementation requires a certain level of organizational maturity to be achieved, establishing a PMO in the form of an EMO is an additional step in the evolution of the organization on the maturity continuum.

The EMO should then be the entity enabling the consolidation of the information (not only data) regarding all sources of resource consumption and revenue generation, with the aim to build, if not be itself, the bridge between the different parts and levels of the organization to propagate the defined strategy. The EMO should also bear the responsibility and the accountability of propagating the "Why" of the organization—its strategic vision—and provide the means to ensure the alignment of all parts and components of the organization with that vision.

It doesn't even necessarily mean that this EMO should replace the PMO. A PMO, defined as being a "project" management office, can co-exist with such an EMO and even be directly connected to it—the EMO being the fuse to prevent the PMO(s) becoming factors of organizational entropy by aligning the monitoring indicators and means with what really provides mastery and visibility and not only an illusion of control.

In their study, Hobbs and Aubry (2010) have established that most PMOs are in fact temporary organizations, and in practice, I have seen the same trend in many companies. PMOs' life expectancy is in fact around three to four years. That short life span is often due to a lack of an established value and benefit statement behind the creation of the PMO. Often, when establishing a PMO, an organization seeks to solve one particular problem encountered within the governance of that organization's projects and programs, and once that problem is solved, the PMO struggles to demonstrate the creation of additional value for the organization and justify its business case, and thus is often dismantled. When a new problem (or sometimes even the same problem) occurs, the opportunity for re-establishing a PMO appears again. The repetition of that cycle leads to a rollercoaster of ups and downs in the level of organizational maturity, but no real improvement.

In a conference paper I presented in 2011 (Lazar, 2011), I proposed to exploit the opportunity coming from that temporality by establishing a formalized PMO Life Cycle, allowing the definition of a continuous delivery of value for the organization. The principle of the PMO Life Cycle is to define phases of the establishment of a PMO, with various roles and responsibilities as well as various placements within the organization of that PMO, depending on the ability of the organization to control its project management processes and the projects and programs within its portfolio (see Figure 3.5).

The first phase, named the "Clean-up," is about directly solving the problem which led to establishing the PMO, such as realigning projects and programs to be back on track. It's often a firefighting phase, requiring both a certain set of skills and competences to be directly and deeply involved within the projects and programs, and also a strong level of authority, which makes that phase often handled by external consultants under the direct authority of the executive level of the organization. As being a firefighting phase, it's often also treated as a project or program, and then by definition it's a temporary endeavor.

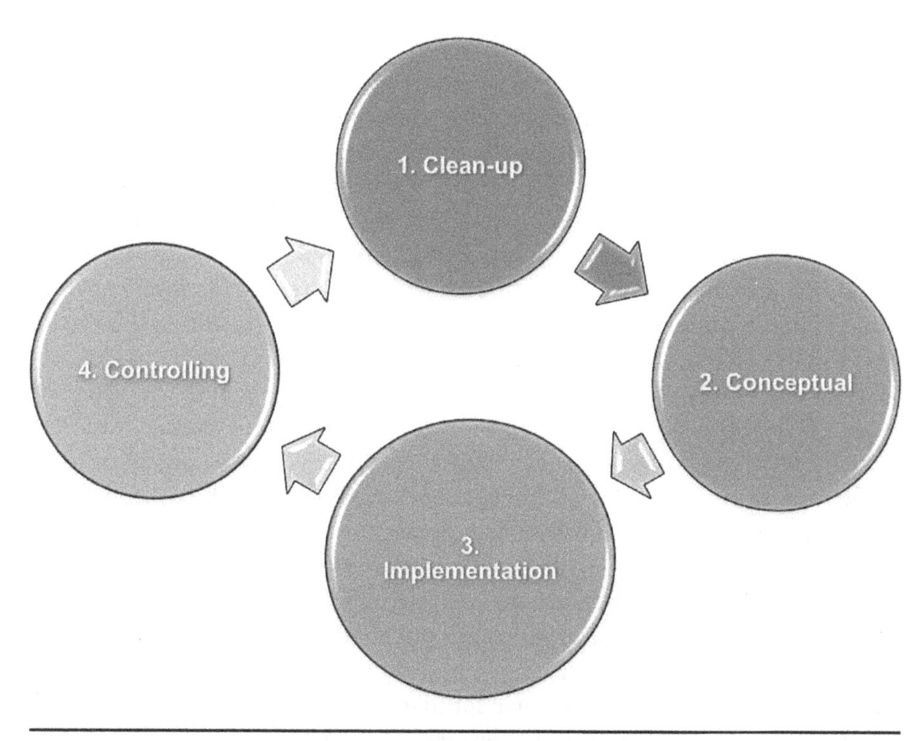

Figure 3.5 The PMO Life Cycle (Reproduced from Lazar, O. Ensure PMO's Sustainability: Make It Temporary! PMI® Global Congress 2011—North America. Dallas, TX, USA: Project Management Institute. © 2011 Olivier Lazar)

The second phase, "Conceptual," starts shortly before the end of the previous stage, at which point it becomes important to understand where the problem raising the need for a PMO came from and how it has been solved by the firefighters. The people and skill set here are totally different. From mercenaries, we now need thinkers, able to analyze the gap between past and present in terms of project management practices, and to derive guiding principles, tools to apply and support these guiding principles and draft a roadmap to propagate these best practices and relevant methodologies and frameworks throughout the organization. Usually this requires a small team, comprising a mix of generally external experts, able to give an external perspective and provide a wide range of project management knowledge and experience, and key internal influencers with a deep knowledge of the organization's business and its culture.

The conceptual endeavor might also be covered by a temporary initiative, project, or part of a wider program put under the authority of an executive level, if not attached to a Quality Management or Operations department or function. There is definitely no need here to establish a separate organizational entity, which would add to the overall complexity.

Once this conceptual phase has delivered its outcome in the form of a validated and proof-tested framework, there arises the necessity to spread the good word and deploy that framework throughout the organization, aligning all projects and programs at each level of the organization with the newly defined framework, developing a common perception, vocabulary, and way of working among the different teams. The "implementation" phase, having different aims, also requires a quite different skill set, even different personalities. The PMO people will need to again get involved within projects and programs, but not with the aim of steering them, but rather training, coaching, and mentoring the teams of those projects and programs to accompany them in the application of the newly defined framework, eventually supporting the whole organization in a wider transformation. I have participated several times in such initiatives (each time run as a program), often having an impact of the organizational structure itself, moving from functional to matrix structures and transforming hierarchical and communication channels. This is where change happens. This is indeed a critical phase; a group of experienced and seasoned experts is required, in addition to which, involving key internal influencers is essential to the success of such an initiative. A clear roadmap for that transformational program, allowing periods of relative stability, will ensure that everyone not only respects the limits of the organization's ability to absorb a certain level of change, but also fosters the chances of that transformation to be successful.

Often, as the transformation progresses, a good practice is to raise the proportion of internal resources involved and decrease the number of external consultants to maximize the knowledge transfer and finally make the change and the new project management framework something really owned by the organization. The key success factor will be to make that change no longer a change. The newly established practices and the eventual new structure, and the implications of that structure on how people work and interact, have to become part of the DNA of the company, and once absorbed become the natural way of doing things. That takes time and effort, which are not to be underestimated, but when properly handled, the benefits in terms of organizational optimization, reduction of entropy and inertia, even overall performance and employee happiness, can be tremendous (Lazar, 2016).

While implementing a new framework and establishing a new organizational structure, there arises a need to ensure that everything works and is used properly, that the teams follow the guidelines and apply the tools, and that the information's reliability is preserved as a key success factor for the implementation of an efficient and effective portfolio management practice. There's then a need for the installment of a controlling body, performing a sort of quality-assurance and data-gathering role, eventually raising alerts if and when a deviation is observed, be it on the performance of the portfolio components or within the application of their governance processes. Establishing that controlling role is the aim of the fourth phase of the PMO Life Cycle, the "Controlling" stage. Again we face a different role of the PMO, requiring a different skill set and even a different mindset. Here we are no longer in the context of a temporary initiative, but rather in a continuous function performed on an ongoing basis. The temptation of establishing a permanent addition to the organizational chart to support this role is often quite strong, but if we look at the competences required here, we can see it can perfectly be assumed by a financial controller, quality auditors, or even better the project and program teams themselves. Building a specific PMO entity at this point is a matter of decision related to the organizational culture, management perspectives and preferences, and of course, a good business case; but it's not always necessary (see Figure 3.6 for the sequence of PMO Life Cycle phases).

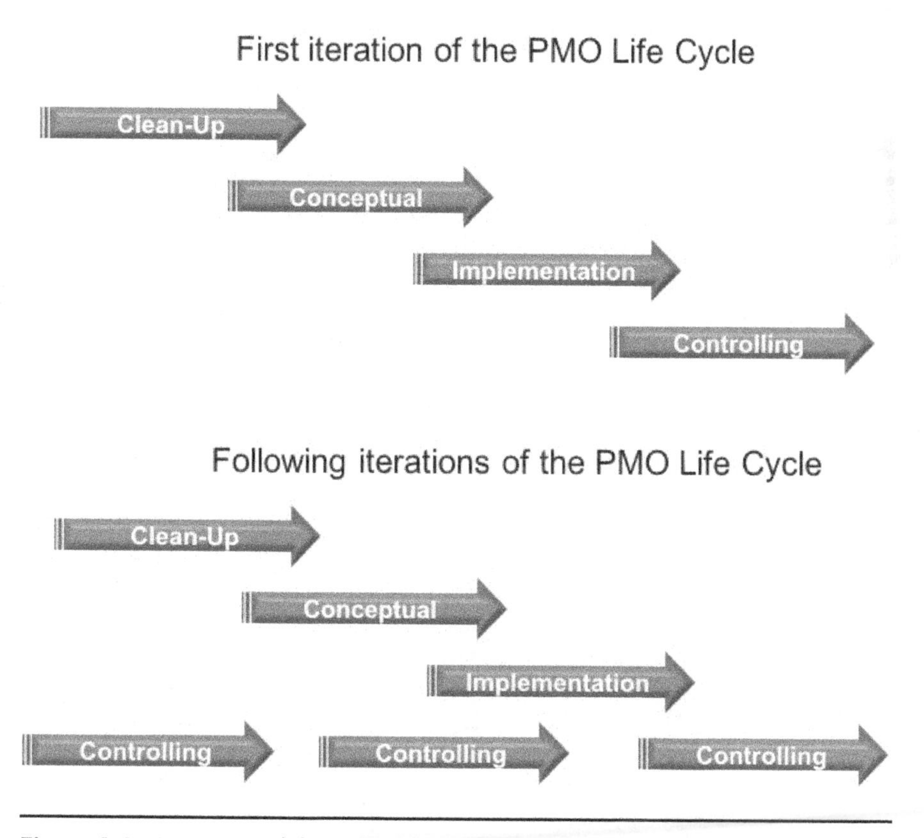

Figure 3.6 Iterations of the PMO Life Cycle

Some of these phases described in the PMO Life Cycle require setting up a formal project or program, but in the end, none of them specifically requires establishing the PMO as a formal entity. What is really important when considering constructing a PMO within the organization is ensuring that the PMO role is continuously assumed somewhere, by someone, not to have a project management office box drawn on the organizational chart. It's also important to be ready to transform that PMO role based on the evolution of the organization's maturity level, and continuously serve as a means to maintain and uphold that level.

As we'll explore more in depth in the following pages, portfolio management is the cherry on the sundae of the project management continuum. It requires a solid foundation of practices and reliable data to be effective and efficient. The PMO is then instrumental in that endeavor, be it formally and structurally established or spread throughout the organization as a role to be assumed by different people at different stages of the evolution of that organization.

Chapter 4

The Second Pillar: Your Organization's Strategy

In the previous chapter, we explored what in fact will sustain and justify the achievement of a proper and established portfolio management practice: organizational agility, or developing the means for the organization to anticipate evolutions and changes within its business environment and adapt to those changes without having to hit the limits of the tolerance to change of that organization, all without losing control over the direction and course of that organization on its way toward the realization of a strategic vision.

The formulation of that strategic vision and then of the related strategy is an essential basis in that effort. But formulating this vision and that strategy is not only about saying, "I want to increase my revenues by *x* percent," or, "I want to be the best at my business." It's even the exact opposite of these statements.

A strategic vision has to be the very justification of the existence of the organization (in whole or part). It needs to be the link that will connect the individual human beings forming that organization (remember that an organization is a gathering of individuals). It has to be the "Why" of that gathering, motivating, giving a sense and a purpose as very well described in Simon Sinek's *Start with Why* (2009).

A strategic vision and the related strategy defined to achieve it should not be formulated in terms of performance, but rather in terms of value. What kind of difference does your organization create? What is the motto behind your actions? What is the impact you want to imprint in your environment?

Performance is a consequence of a successfully defined, formulated, and implemented strategy, not an aim in itself. As Jack Welch said in 2009, quoted in an article by Steve Tobak:

"'On the face of it, shareholder value is the dumbest idea in the world,' Welch said. 'Shareholder value is a result, not a strategy . . . Your main constituencies are your employees, your customers and your products'" (Tobak, 2009).

And indeed, that strategy should be defined within a certain timeframe. The strategic vision can be non-temporal, but the strategy itself as being a plan should have a timely target.

4.1 Defining Your Strategic Horizon

As portfolio management is time framed and is a direct reflection of the strategy, we need to define a temporal window within which our decisions will be implemented, and our initiatives will justify the value they create for our stakeholders. That "Strategic Horizon" we have to define is not coming from nowhere, it's merely a decision—more of a fact, a parameter which has to take into account a certain number of parameters. Among these parameters, there is one essential one which we have already approached: *organizational inertia*.

4.1.1 Measuring Organizational Inertia

As defined here, organizational inertia is the lag in the implementation of strategic decisions in the organization—in other words, the time it takes for a decision made at the executive level of the organization to cascade down the organizational chart, start to be implemented, and the feedback about that implementation to go back up the ladder to the executives. That inertia is of course influenced by various factors.

First, the organizational structure. It's quite obvious that a flat structure, such as a projectized or a strong matrix organization, generates less inertia than a functional one. Second, the entropy of the organization, that lag generated by the complexity of its governance structure, its processes, systems, guidelines, reporting, and measurement systems. All these things, necessary or not (often not), slow down the decision-making and decision-executing processes by adding layers of validation and measurements, often inducing noise in the transmission of the information within the organization. Of course, to add to the contributors to organizational inertia, the very culture of the organization has to be taken into account. Cultures of open and direct communication will generate less inertia than the ones where, by principle, communication is very structured and formal. Also, the business environment will influence the level of inertia: The more the environment will be subject to regulations and legal constrains, the more it will induce inertia and eventually create entropy.

Organizational inertia is then something we can measure. It's usually expressed in time (years, months, weeks), considering the time a decision takes to go through the organization. That's the value we'll account as being "i."

It's not rare in many organizations to see a level of inertia around two to three years, sometimes more. It might seem high, but in fact that level of inertia taken in isolation, out of any context, is meaningless. It depends on so many factors and is relative to so many parameters that a certain level of inertia that might seem very high for a small hi-tech start-up will be perfectly normal for a big pharmaceutical company evolving in a highly regulated environment.

It can be interesting to measure inertia itself though, if used as a benchmark among departments within a single company, or among organizations from the same industry. It is obvious that you will want to align the levels of inertia within your company to the department having the lowest one. The whole organization is as fast as its slowest part.

Organizational inertia can also be used as an indicator to measure the evolution of organizational maturity as you progress with the implementation of portfolio management principles, not as being a targeted absolute value in itself, but rather as a percentage of reduction of that inertia. While gaining organizational agility, reducing entropy, which are both important outcomes of what I try to explain in this book, you would need to demonstrate the gains and benefits you have achieved through the efforts invested into the organizational transformation implied by the development and implementation of portfolio management. Organizational agility and organizational entropy are both concepts. It's difficult to measure concepts. But organizational inertia is something tangible which we can measure; and if we can prove we have decreased it, it's a tangible outcome and demonstrator of success.

Organizational inertia is also a component of what defines your company's competitive advantage. Companies with a lower level of inertia than their competitors are capable of quicker decision making and decision executing, can react faster to changes in their environment, and respond faster to the evolution of their customers' demands.

Then whatever level reaches your "i," don't panic—it's just a challenge to improve and an important input to determine how far ahead you should plan your organization's strategy. Because you can, and you will have to improve that level of inertia, that's exactly the aim of this book. Just be conscious that you can't *eliminate* inertia. It's a physical phenomenon inherent to the very human nature of the concept of organizations.

4.1.2 Calculating Your Strategic Horizon

Given that measured level of inertia, we need to take into account another factor related directly to the nature of your company's business. That factor is the so-called *business environment evolution rate.*

When taken at the level of a whole business environment, this represents also how fast or how often you have to launch a new product or service to remain competitive in your market. It represents, in fact, the dynamic of your business environment, how often your surrounding environment changes. By the way, these changes are not necessarily new products and services released by your competitors; they can come from newcomers disturbing your business model (what we use to call *uberization*), changes in the regulations or political and economic environment, or even coming from your own organization through the execution of your strategy and the realization of your strategic vision.

How dynamic your environment is can be measured by an assessment that I will not discuss in these pages. That's a topic for market analysts, strategic consultancies, and other related experts. Let's consider it herein as an input taken from a deep and seriously conducted market analysis.

That evolution rate will of course also vary from one market to the other. Definitely, the evolution rate of the telecom market is higher than that in aeronautics or automotive, which are in turn higher than in pharmaceuticals or fundamental physics.

Then, what will be the evolution rate of your market environment? Do you have to release a novelty every six month, every 12 months? Let's designate that number as "Er."

What do we do with these parameters now?

Let's take a practical example. Imagine your company navigates within a market environ whose evolution rate is 12 months (Er = 12). And let's imagine that your company's level of inertia is 18 months (i = 18).

How far would you have to look forward in order to avoid a catastrophe?

Would 12 months be sufficient? No, because it takes you 12 months to implement any change within your strategic course, so you would see the threat coming six months too late.

How about 18 months? Not sufficient either. By looking only as far as your level of inertia, you place your organization in a constant (and not continuous) changing course. When a change has just been implemented, it is already time to start initiating a new one, which will also become obsolete as soon it is put in place. You'll in fact place your organization, and thus the people who are composing this organization, in a constant panic mode, running in a spinning wheel. You will very quickly hit the limit of your organization to absorb a certain level of change, trigger resistance at the best; the worst-case scenario will be a total loss of control, as the monitoring tools developed in a given situation are not valid anymore to face the one to come.

We need then to expand our strategic horizon beyond these two limits.

At least double their sum.

The formula used to calculate how far we need to put our strategic horizon (SH) would then be the following: SH = 2 (i + Er) (see Figure 4.1).

$$SH = 2(i + Er)$$

Eg: i = 12, Market Evolution Rate (Er) = 6
Strategic Horizon = 36 months

Figure 4.1 Calculating Your Organization's Strategic Horizon

To take our example, the strategic horizon of that company would be as follows: 2(12 + 6) = 36 months, or 3 years.

To be sustainable and manageable, that company should develop a strategy with a horizon of three years ahead.

Picture yourself while driving your car. Where do you look? As far as possible or just at the edge of the hood? As far as possible of course, to be able to anticipate any incoming threat and introduce just a slight change on the steering wheel rather than introducing a large and quick turn which would put you at risk of losing control of your car. It's exactly the same principle which is applied here. And if you pay attention, you'll notice that you make a lot of changes in direction that you don't even consciously notice to keep your car on track.

Actually, that's what organizational agility is about. Being able to anticipate the modifications of our environment to adapt our actions soon enough within the limits of our ability to absorb a given level of change.

What do we do with that? Why is it so important?

Positioning the strategic horizon plays a key role in the elaboration of the overall strategy of organization, and from that in the construction of our portfolio(s), as the portfolio represents exactly the concrete realization of that strategy.

The strategic horizon will determine how far we will plan our projects, programs, and initiatives, decide which ones to launch or not.

When forecasting our resource allocation, we will push these forecasts and estimates up to the strategic horizon.

When prioritizing our portfolio components, we will favor components delivering benefits and profits within the timeframe of the strategic horizon. In our example above, programs will have to create benefits within the next five years, and the assets created by projects should have a payback period shorter than five years. The net present value (NPV) of these assets should also not drop to zero in a shorter time than five years.

As sometimes happens, calculating this strategic horizon will push us to extend our forecast exercise, and it might also lead us to limit that forecast. Making forecasts beyond that horizon would at best present a too high level of uncertainty, or too low precision, to be of any relevance, or at least represent a level of effort too high compared to the benefit of making estimates further than the strategic horizon.

Determining the strategic horizon tells us how far we have to look, but also how far we are capable to look, sizing the forecasting effort to its best cost/benefits/accuracy ratio.

Of course, as portfolio management is an ongoing, continuous, and iterative exercise, often based on a yearly calendar, each year we'll push forward that strategic horizon by one year, but each year at least, reconsidering the relevance of that calculation. Maybe the dynamic of your business environment has changed, maybe you have been successful in optimizing the level of inertia in your organization, gaining then in reactivity.

4.2 Constructing Your Strategic Vision

Your company's strategy should be framed by your strategic horizon, but that's not necessarily true about your strategic vision. That vision, which will give a sense and goal to the strategy, justifying the investments made by their contribution into realizing it, is what has to be the driver of the people within the organization. Clear, ambitious, inspiring, and motivating. One of these goals for which people will go the extra mile . . . Yes, indeed . . . But in fact, formulating a strategic vision is not an easy exercise.

How many times have we heard statements such as "my strategy is to increase market shares/revenues/profits by *x* percent," "my strategy is to be the best," or "my strategy is to reduce our costs by *x* percent," and so on . . . ? Too many indeed. These statements are exactly the opposite of a strategic vision. A strategic vision must be the expression of the rationale for the existence of the organization, its "Why."

According to a definition of strategy that I like a lot, it is, ". . . the process by which an organization envisions its future and develops the necessary procedures to achieve that objective" (Pfeiffer, Goodstein, & Nolan, 1986). Strategy requires a vision—vision is not strategy, and strategy is not vision.

So, if the quantitative terms mentioned above are neither a strategy nor a vision, how should we formulate it?

4.2.1 *Aiming at Value*

Actually, the rationale of the existence of your organization should be described as its impact on the environment, on society—the track left behind it. Which, speaking of a track, is exactly what will make people follow you and your organization; that's what attracts and retains employees and clients.

What if . . . we would define the strategic vision as being the expression of the desired configuration of the organization's market or environment in a defined future?

It would then describe the status of the business environment on which your organization navigates once being influenced, or even shaped, by it. How would you want your environment to look like in the more or less near future?

Here's an example: "To organize the world's information and make it universally accessible and useful." That's Google's vision statement. It doesn't speak about the company itself, what it should look like. It doesn't even speak about products or services, nor does it give any timeframe. It describes an environment, business or societal, as the executives of the company see it in a more or less distant future. The company is the tool to shape that future to come.

The benefit of stating the vision in qualitative, value-oriented terms is to trigger an emotional response, enabling adherence, motivation, and engagement. It's also a goal which leaves enough of freedom and space to define the strategy according to the evolution of the business environment, allowing us eventually to adjust the strategy without having to make much compromise on the ultimate goal, keeping the aim in sight and finding ways to reach it.

These ways to invent, this route to draw, in order to reach that goal is what will constitute the organization's strategy, including the definition of the means necessary to achieve it, the actions to undertake to exploit those means, and the plan describing how and when at least a piece of the vision has to be completed.

4.2.2 *Strategy and Governance*

There are then two aspects, or dimensions, to consider—the strategy itself, as described above, being the set of initiatives (projects and programs) that will create the means to achieve the organization's ultimate goals, and the operations (business as usual) exploiting that means to effectively achieve these goals.

If we would give a definition of strategy, it could be the set of programs, portfolios, projects, and operations undertaken to realize the strategic vision of the organization (Lazar, 2010).

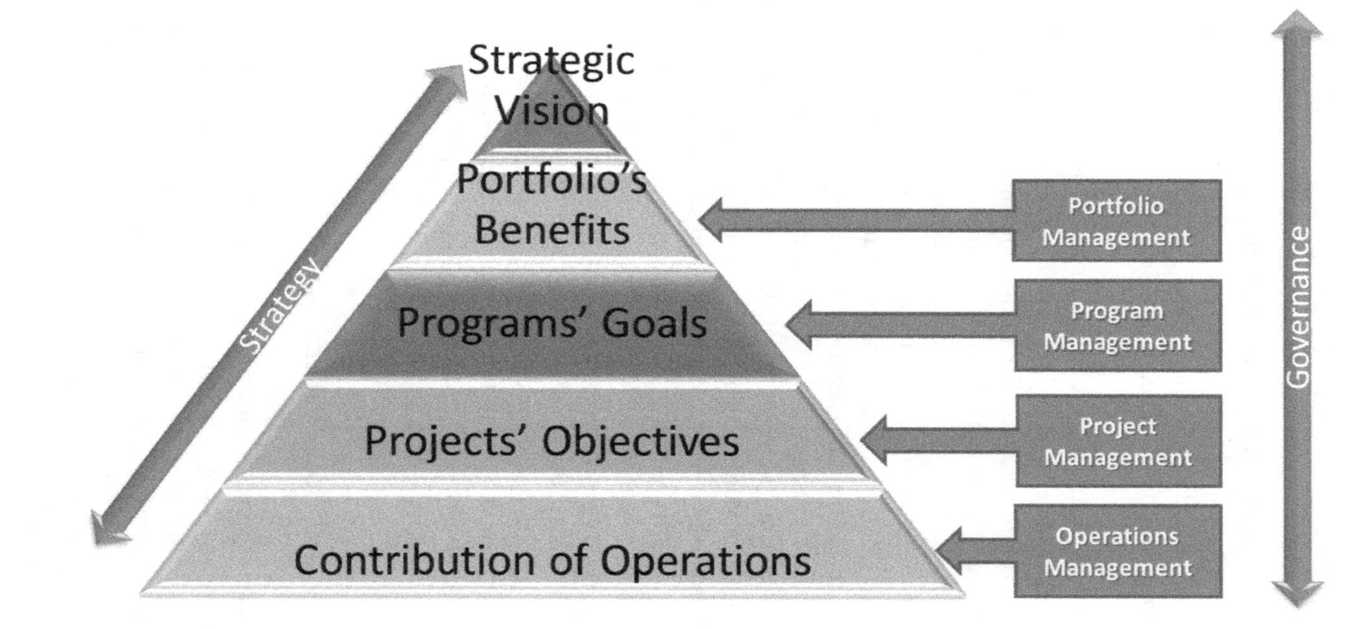

Figure 4.2 Strategy and Governance: Two Sides of One Coin (Reproduced from Lazar, O. The Project Driven Strategic Chain. PMI® Global Congress. Washington, DC: Project Management Institute. © 2010 Olivier Lazar)

To drive these initiatives and operations, the organization will have to put in place a certain set of processes, tools, techniques, guidelines, etc. This is what constitutes governance.

Organizational governance is the set of processes, tools, methods, and controls applied to conduct the realization of the strategy (Lazar, 2010; see Figure 4.2).

Strategy and governance are then two sides of the same coin. Strategy describes how the organizational goals will be achieved, and governance provides the means to execute and monitor that execution through the different process levels integrated with each other, such as project, program, and portfolio management.

The construction of the strategy is of course essential, starting with the strategic vision, but the efficiency and effectiveness of any organization lies within the ability of that organization to establish the proper governance model. Governance is what creates inertia and entropy; governance is what triggers resistance to change. The root cause of failure is rarely the strategy itself, but rather the absence of it. Defining the strategy, formulating it from the expression of the strategic vision, is in fact a governance process.

4.2.3 Organizational Structures and Portfolio Structure: The Dog or the Tail?

The organizational structure in place in your organization is not really a matter of choice. It depends on how business is conducted within that organization—its culture, its size, etc.

That structure and how it impacts the flows of authority, decision making, communication, reporting, and delivery will of course also impact the way you will have to structure the portfolio itself (or the portfolios themselves as there can, and probably will, be several portfolios defined within the organization, and even portfolios of portfolios).

In the case of a functional organization, the portfolio structure is simple. It fits the organizational one, but you will have at least two levels of portfolios:

- The global overall portfolio, consolidating all activities of the organization, including all initiatives (projects and programs) and operations. This global portfolio will be directly under the supervision and control of the executive level of the organization, and in many cases the role will be held by the CEO.
- The main structure will be based on the functions, defining one portfolio per department or branch.

In the functional organization, operations are at the core of its activities, so they will also be at the core of the portfolio management exercise. The portfolio management role will then be held by the functional manager. It will lead to a vertical portfolio structure that is mainly a financial consolidation of ongoing operations and intra–business-unit initiatives. The main purpose will be to monitor the consumption of resources related to the generation of revenue and profit and being able to forecast and allocate these resources on a yearly basis to the different operations' streams, up to the strategic horizon, which can be far away in time or short-term. The strategy here is quite linear and doesn't vary much over time, so the importance of positioning at a significant and relevant distance in time becomes less important.

If the organizational structure is based on the projectized model, it's even simpler than in the case of a functional structure. There's a single portfolio, with almost no structure at all—a kind of "big basket" consisting of all projects running in the organization, directly under the supervision and authority of the executive level. The aim here is to allocate the resources to the more profitable projects and balance their cost with the generated profit. Often, these organizations apply a simplistic version of portfolio management, and their strategy is limited to maximizing profits and responding to clients' demands as well as they can, as fast as they can. But as opposed to the functional organization, here, positioning the strategic horizon as far as you can takes on a certain importance, because these organizations need to anticipate as much as possible the variations of their market and the demands of their clients, being as agile as possible to adapt to these evolutions within their environment. It's a matter of survival. "Change or Die," as stated to me by Troy Hazard, a world-renowned business leader and author, during an informal discussion we had back in 2010.

Structuring the portfolio will become far more complex when it comes to the matrix organizational structure. As we have already detailed in the previous pages, we can face two different sorts of matrixes:

- The *weak* matrix, in which the projects and programs undertaken within the organization serve the operations by developing the business capabilities that the operations will exploit to generate revenue and profit.
- The *strong* matrix, which works the other way around—that is, the organizational functions serve the projects and programs by providing them with competent, motivated, engaged, and performant resources, thus having the majority of the revenues and profits be generated by these projects and programs.

I won't address the so-called *balanced matrix,* as this model is definitely one that, even if existing, leads to catastrophic situations and is clearly to be avoided at any cost. When the authority is equally spread out, there's no authority anywhere.

The portfolio structure within a weak matrix organization will be similar to the one we can see in a functional one. The integration of operations is vertical; they belong to their specific function, as is often the case with the capacity development projects and programs that are attached to a specific department. Of course, these organizations will also initiate transversal projects and programs, but then these will be attached to the executive portfolio at the top level of the organization. Figure 4.3 shows an example of a portfolio structure within a pharmaceutical company. This company is divided into business units, in pharma often called *therapeutic areas* (or TAs). Each of these TAs is dedicated to a specific kind of treatment or type of disease (oncology, fertility, neurodegenerative diseases, etc.), wherein different drug development programs are launched, in the hope of developing a new medicine to be sold on the shelves of pharmacies and hospitals. Within these programs, different projects are undertaken, all of which contribute to the development of a new capability for the organization and to putting a new product on the market. These projects can be of various kinds and with different aims but are all within a program, helping to achieve

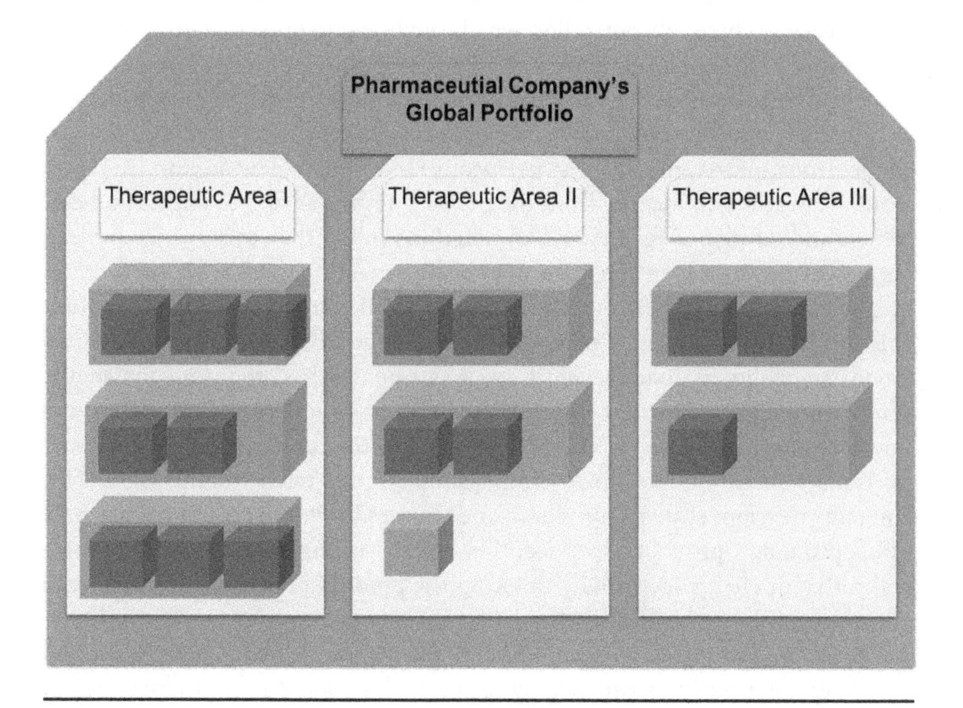

Figure 4.3 Example of a Portfolio Structure in a Weak Matrix Organization: The Pharmaceutical Company

the objectives of that very program. Sometimes, a few projects can be developed outside of a specific program, as they contribute to several drug development programs, such as a project to develop a new medical device—say, an injector which can be used for several different drugs.

If one of the projects within a drug development program fails—for instance, a clinical study shows negative results—it puts the entire program in jeopardy; but if one drug development program has to be stopped, it doesn't mean that the other drug development programs within that therapeutic area will have to be stopped too. They are not interdependent in their success, nor do they necessarily contribute to one another's objectives. They are separate objects, or components, but they all contribute to a consolidated business performance and often use similar resources taken from the same pools. Each TA is then a portfolio, if we apply the classic definition of a portfolio as stated in the first pages of this book.

On the top level of the organization, the different TAs are consolidated within the organization's global portfolio, to which they are sub-portfolios as part of the whole. If a PMO exists within that organization, it will often be positioned at that top level and will have an overview of the global portfolio, even if subsets of that global PMO can exist within each TA to monitor more closely each of these sub-portfolios, which can be a good practice to recommend in terms of PMO construction.

The structure of the portfolio in a strong matrix organization is not necessarily very different from that of one in a weak matrix organization. But in the case of the strong matrix, the business perspective is different. The organization generates its profits and income from the projects' results as sold to clients and customers and developed for them. The function's role is then to serve the projects (as opposed to its role in the weak matrix) by providing them with appropriate, competent, and motivated resources. The portfolio structure must then reflect this project-driven business perspective and essentially give a sort of "product-based" portfolio approach, even if one way of entry into the portfolio should provide the resource owners with the necessary visibility over where, when, and how their specific resources are used now and in the future. Figure 4.4 shows a portfolio structure established within such a strong matrix organization. This company develops systems, products, and services for its clients. Once they have developed a new product or service, they tend to sell its implementation to several different clients, eventually tailoring the product in accordance with each client's specific needs. That's where their income comes from—their ability to sell their products and services and the related maintenance activities—and they keep their business running by regularly developing new products to stay current on the market. They are active in a quite dynamic market, in which the technology evolves fast, and they need to sustain a continuous stream of new products and services.

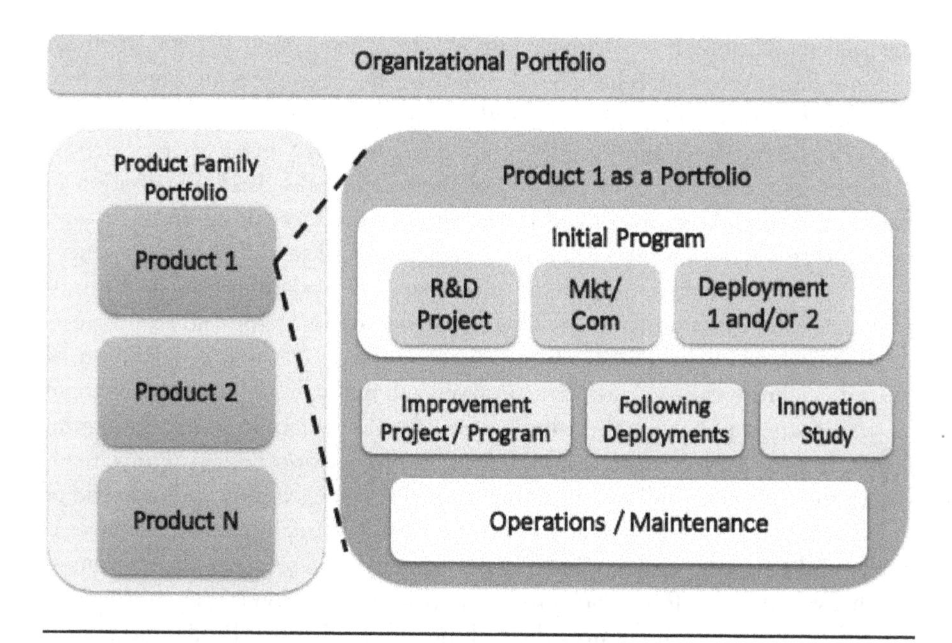

Figure 4.4 Product-Based Portfolio Structure in a Strong Matrix Organization

The business model of such a company is quite simple: They develop products, which generates their income. For each new product, they launch an initiative that aims at developing a new capability for the organization, which consists of putting a new product on the market. That initiative includes a research and development project, a marketing development project, often one test deployment project conducted for a lucky client designated as the lab rat, and sometimes a second or even a third test deployment to make sure the success of the first one was not a happy coincidence.

All these actions, which have their own specific results to produce within a defined timeframe and with definite resources, are then projects. These projects contribute all together to achieve an expected business benefit, which is developing the organizational capability to put a new product on the market. The sum of these projects represents a program in which all components are interdependent in their contribution to a common purpose: an initial development program that is considered ready to be closed when all checkboxes indicating that the organization has a new product ready to be marketed are ticked.

Once the initial development program is completed, the company sells that new product to clients, during which they launch implementation projects ("projects," because they have a concrete, tangible deliverable; specified resources; and a definite timeframe)—as many as they can, as often as they can. But if the implementation project with client A is a success, it doesn't mean that the implementation project with client B will be as successful. There is no direct and immediate correlation between the successes of the implementation projects, nor does the failure of one necessarily affect the existence of any of the others.

We're definitely facing not a program structure here, but a set of independent projects. The only common aim of these projects is their contribution to the same revenue account linked to the specific product. In addition to that, each implementation project eventually triggers a set of so-called "maintenance" activities, which continuously require propagating updates, fixes, and minor adjustments to the product itself. These activities are on-going and repetitive—small iterations of similar actions perfectly fitting within the definition of operations. From time to time, the company will make more important changes to the product, requiring the launch of a specific project or even a program, depending on the depth and complexity of the change or product improvement.

All these elements follow a typical product life cycle from inception to end of life, representing for the company a business stream of income, spending, investments, and a node to measure financial performance and return on investment—in other words, a portfolio. Each product in this organization is then managed as a portfolio by an identified individual whose official title was Product Manager, who defines the product's strategy and is accountable for the performance generated by that product, but whose role is now Portfolio Manager, defining that strategy, translating it into tangible outcomes and performance objectives, and allocating organizational resources among a set of prioritized components (projects, programs, and operations), depending on their contribution to the product's strategy and its expected performance, all aimed at maximizing that performance by optimizing the usage of organizational resources.

And that product portfolio is one among several within a business unit—a gathering of similar products. That business unit's aim is to consolidate investments and revenues generated by a family of products; it's then also a portfolio, as the business unit manager manages a portfolio of portfolios. And that business unit is part of an organization with several business units under the supervision of a global manager, who has to make sure that the set of business units generates enough performance to sustain the existence of the organization by managing a portfolio of portfolios of portfolios . . . and guess what, in this real-case example, the organization I mention belongs to a larger group, headed by a group of CxO's who then have to manage a portfolio of portfolios of portfolios of portfolios . . . the very principle of Russian dolls.

Access to the portfolio data will mainly be exerted from the perspective of the projects and programs, with a focus on products rather than on resources per functions. Of course, an entry such as in the weak-matrix–related portfolio structure—allowing the resource owners to see how much of their resources are now and will be used on which project—is important.

The definition of the portfolio structure is an important step in the set-up of portfolio management within an organization. It conditions how the strategy will be executed and how the decisions will be made. Putting the focus on functions or on projects creates a different approach to authority, prioritization, and measuring performance.

In a functional or weak matrix, in which the business is driven by operations, the functions will hold the keys to decision making. In a projectized and strong matrix, the projects and programs will be at the core of that decision making. This factor will also greatly affect the formulation of the strategy. And it's not only a matter of choice; it's more of a fact—a parameter to take into account, an "Enterprise Environmental Factor," as the Project Management Institute calls it in the *PMBOK® Guide* (2017a).

4.3 Constructing Your Portfolio's Strategy: Building the Opportunity Chain

As the portfolio of the organization represents and reflects its strategy, it becomes the tool to communicate that strategy throughout all the layers of the organization, translating that fuzzy and sometimes undefined vision into a set of concrete and tangible actions—all specific, measurable, achievable, relevant, and time bounded. Wishful thinking, some might say . . . but portfolio management, properly used and carefully constructed, can even accomplish that—the translation of the strategic vision into the strategy itself.

What is the aim of any strategy? Exploiting business opportunities, indeed. To identify which opportunities and how to exploit them, we'll construct the backbone of the strategy, which will be the foundation of the expression of the organization's portfolio, what I call here the *opportunity chain*.

"Opportunity" because we'll look at maximizing value for the organization's stakeholders, trying to develop the most rewarding strategies to achieve the vision and maximize the organization's sustainability; "chain" because it forms the tools by which we will tie together the different layers within the organization, connecting the dots and aligning these layers on a common and shared perception of the objectives to achieve and the vision that drives the organization.

4.3.1 Stating the Initial Concept

In Section 4.2, we defined *strategic vision* as a targeted configuration of the organization's business environment, setting up the organization as the tool to shape that environment.

The competitive differentiator will reside in the value created for the organization's stakeholders and the ability to sustain it over a defined timeframe, itself defined by the strategic horizon mentioned earlier.

But what is value? What does it mean to create value?

As described in his book, *A Framework for Value Management Practice—2nd Edition*, by my friend and colleague Michel Thiry (2013), value is a perception—a perception from the point of view of your stakeholders. What's a stakeholder? It's any individual, or group of individuals, directly or indirectly having an impact, or being impacted by, your initiatives, their processes, or their results. Quite a wide range of people.

But value is also a balance, an equilibrium between the benefits we generate and the resources we use to do so. And in terms of portfolio management, it's the core concept of what we'll be aiming at, what we will in fact be, managing.

Thiry states that the value balance is a bit more than a simple balance between benefits and resources. There is a balance to be found within each side of the equation. On the benefit side, there is a balance to achieve between the benefits expected by our stakeholder(s) and the ones which we will effectively generate. If our stakeholders express 150 expectations with regard to a particular initiative we want to undertake, of course we won't satisfy all of them, because we probably won't have sufficient resources to do so, and certainly because some of these expectations will be either in contradiction to our leading vision or in contradiction with each other (if I do A, I can't do B; if I do B, I can't do C; if I do C, I can't do A). But if you only satisfy two of their expectations among 150, you certainly won't create value, you'll create frustration instead.

On the side of resources, there's a balance to be found as well. Indeed, by defining the benefits to generate, we'll be able to estimate the necessary level of effort to dedicate to achieving those benefits—the required resources. But these required resources have to be balanced against our capability. Will we invest our full capability into a single initiative? Probably not. And by the way, it's one of the aims of portfolio management to find this balance and, from determining our capability, define how much of this capability is available to cover each specific initiative or portfolio component according to its relative level of priority. As Thiry says, we can claim to have generated value if, and only if, all these aspects are in balance, when the ratio between stakeholders' expectations and covered objectives is equal to 1, when the ratio between required resources and organizational capability is equal to 1, and finally when the ratio between resources and benefits is equal to 1 (see Figure 4.5).

Value bears different characteristics. It is subjective: Each one of your stakeholders will have a different perception of the value generated by your initiatives. Even the value represented by this book is relative; some readers will go through it simply because they're interested by the topic, others because they

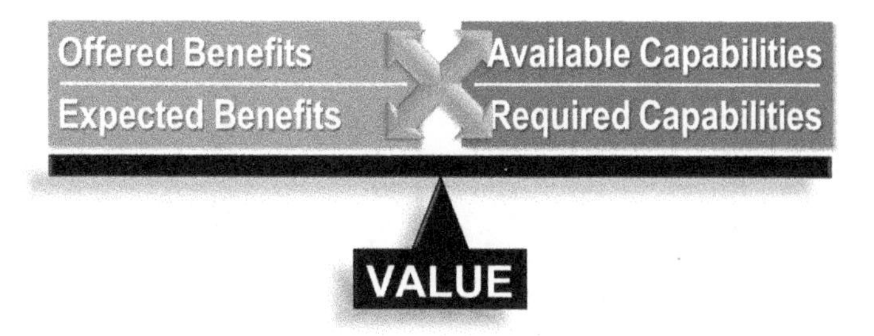

Figure 4.5 Value Balance (Reproduced from Thiry, M. *Program Management* [2nd ed.]. Abington, Oxon, UK: Routledge/Taylor & Francis Group, with permission. © 2016 Taylor & Francis Group)

face a specific problem within their portfolio management practice and expect to find the solution, others because they are preparing a portfolio management professional certification, etc. The same product will be perceived differently depending on the specific needs of each reader.

Value is relative, but relative to what? To our ability to create it. If we can't afford to launch a certain initiative, there's little chance we can generate the expected benefits related to that initiative.

And finally, value is (or has to be) measurable. Basically, if you can't measure something, it probably means that that thing does not exist. It's like temperature. What would be the temperature of a total, absolute vacuum? Zero Kelvin? –273.15°C? –459.67°F? No. There would simply be no temperature to measure. As found on Wikipedia (2018, February 11), "Temperature is a proportional measure of the average translational kinetic energy of the random motions of the constituent microscopic particles in a system." In a total vacuum, there would be no particles, hence nothing to measure. With value, it's the same.

Value is also a construction. It's the result of an evolutionary process.

It starts with an idea, an initial trigger, the basic concept of your initiative. But there's no value in an idea. You might be willing to do many things; a lot of nice ideas can cross your mind, but that it doesn't mean you realize them, and there's no value in an unrealized idea. In order to move forward in the process of value creation, we need to be able to transform our idea, our concept, into a decision, saying, "Yes, let's do it!" But is there any value in a decision? Actually, not much. You can decide many things as well, but that doesn't mean you will realize anything and create any value. As the five frogs story demonstrates, "Deciding is different than doing." To move forward toward the creation of value, we need to be able to transform our decisions into action—in other words, implementing, executing, deploying, realizing, and producing a tangible result. That transformation of a decision into action requires us to have

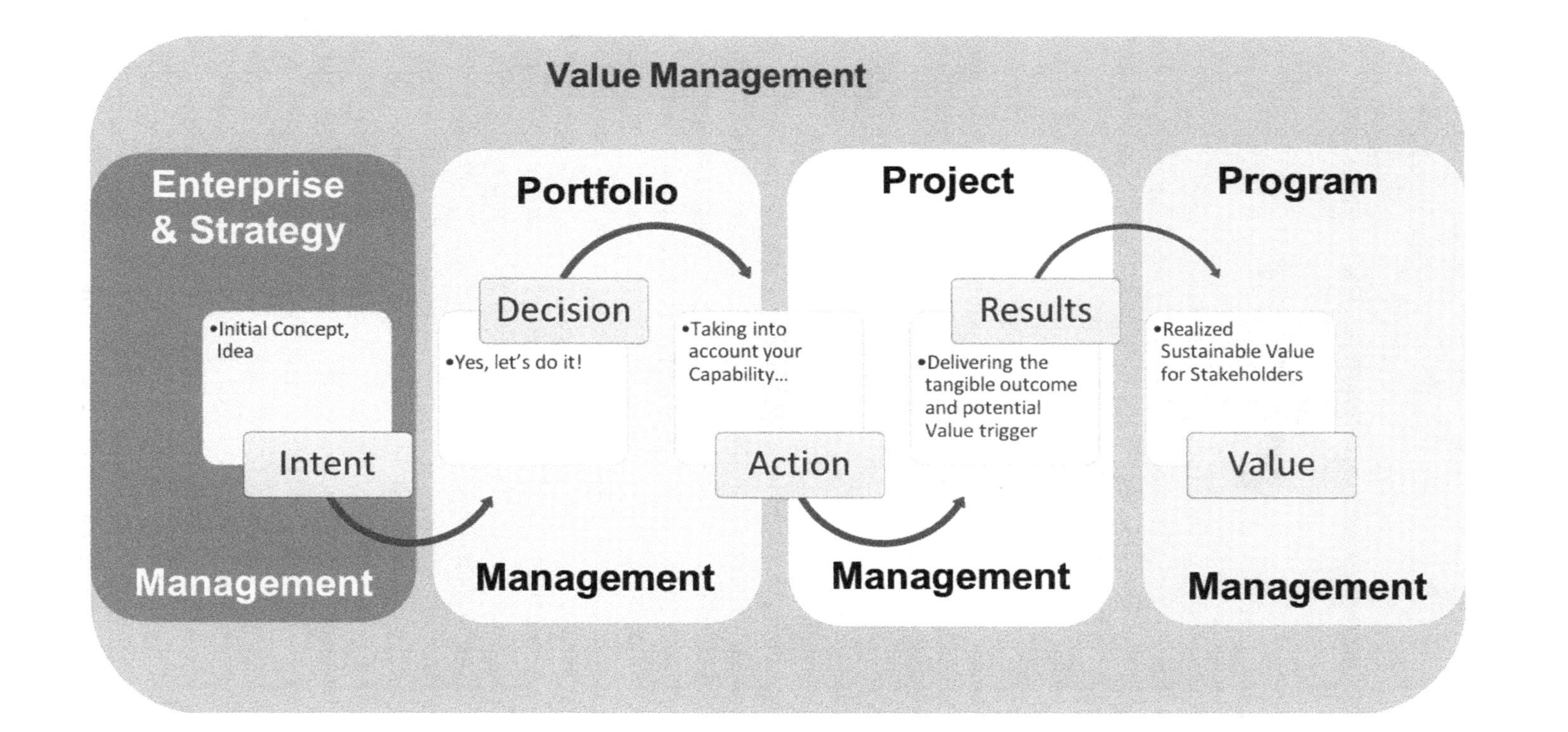

Figure 4.6 The Construction of Value (Reproduced from Lazar, O. "Aligning the Organization through Portfolio Management"; https://learning.pmi.org/in-person-course/747/51/aligning-the-organization-through-portfolio-management. © 2015 Olivier Lazar)

the necessary resources available. That's exactly where portfolio management comes into the picture within an organization. Portfolio management is the governance layer providing us with the visibility over our capabilities, our ability to transform our decisions into action, telling us if and what we can afford to trigger as value-creation–oriented actions.

This is also where project management appears. Project management is what actually transforms decisions into action and produces the tangible result. But is there any value in a tangible result once I have obtained it? Almost, but not yet . . . To generate value, the result needs to be integrated within the organization, a majority of your stakeholders must perceive it as such, and it has to last over time. A result generating a one-shot value, time limited and perceived as such by only a few of your stakeholders, is no value at all. Perception of the benefits generated by the outcome of your initiatives—your results—must be shared by most of your stakeholders and last at least up to your strategic horizon.

And this is where, in an organization, program management appears. Program management integrates your tangible outcomes within the organization, generates change, allows obtaining benefits, and creates value by transitioning the results to operations, where they become not only value triggers but value generators (Lazar, 2015c).

Value is then an integrated organizational construction, combining inputs and outputs from strategy, portfolio management, program and project management, and operations (see Figure 4.6).

The value construction process starts then with the definition and expression of your initial intent: the very root of your aim, the "Why?" in the sense we mentioned earlier when discussing Simon Sinek's *Start with Why* concept (2009). That initial trigger, the strategic trigger of our portfolio, will constitute the foundation on which we'll base the construction of our opportunity chain.

4.3.2 Who Are Your Stakeholders? How Do They Feel? The 4i's Model

As the very definition of value comes from our stakeholders, we need to know them. Starting from our initial trigger, we'll have to determine who will be these individuals or groups of individuals concerned by our portfolio's initiative. Depending on the layer within the organization at which this stakeholders' identification process takes place, the group can be more or less precisely defined, and more or less detailed to the individual level. It's a rather simple exercise consisting in interviews, market analysis, brainstorming, and data gathering, from where we'll consolidate a list of any concerned stakeholder. That list has to be as exhaustive as possible. A forgotten stakeholder might become a very present threat if not addressed, taken care of, and anticipated in its influences and impacts.

Stakeholder management is an essential part of organizational management, at any layer, strategic, portfolio, program, and project. It covers the political aspects at stake within any group of humans, then within any organization.

Following this identification, we also need to know how they feel (literally) and how they position themselves with regard to our initial intent as the basis of the to-be-defined portfolio. We will expand an analysis and classification model made popular by Johnson and Scholes (1997) with the basic concept of the influence grid, but I've added a supplemental dimension to it. We'll use it as the "4i's" model, based on three parameters:

- **The level of Influence of the stakeholder.** The more a stakeholder can impact our initiatives, the more influence that stakeholder will exert.
- **The level of Interest of the stakeholder.** The more a stakeholder is impacted by our initiatives, the more interest that stakeholder will demonstrate.
- **The Intent,** representing the nature or the orientation of that impact. If it is positive, we'll have a supportive stakeholder. If it is negative, we'll face an opponent. Or it can be neutral, neither positive or negative.
- **The Importance of the stakeholder from your specific perspective.**

The 4i's model allows us to identify four main categories of stakeholders:

- First, **the Key-Players.** High level of Influence and high level of Interest. These are the people who will be the closest to our initiatives, deeply concerned and involved. They are the ones with whom we'll have to interact on a daily basis.
- **The Marginal,** on the opposite side of the model. Low level of Influence and low level of Interest. They are the furthest away from us. We'll still need to keep an eye on them and their evolution. Depending on the changes occurring during the execution of our plans and the delivery of our various portfolio components, they might change in position, and we might even want them to change position on this influence grid.
- **The Influential,** having a high level of Influence, but a low level of Interest; they are the ones who we need absolutely, but they don't need us. It's a very important category of stakeholders, as blocking points in the realization of our portfolio components can come from them. If they are not satisfied, they potentially have the power to stop any of our initiatives, if not to kill it. And it won't change their lives . . .
- Finally, **the Affected.** High level of Interest, but low level of Influence. They are the victims of the change, they are the users of our systems,

these are the ones who will have to exploit the results we'll produce. But they are also the ones who transform these results into value or into a waste. The affected stakeholders exert resistance to change. And that resistance comes from fear. They are the ones who decide on the absorption of a certain degree of change. They generate organizational inertia, as mentioned in previous chapters. In fact, they are the most important among all four categories.

And of course, any stakeholder belonging to any of these categories can be either positive, negative, or neutral.

This piece of stakeholder analysis is essential to any initiative (see Figure 4.7). It will feed an entire set of future developments, such as the communications

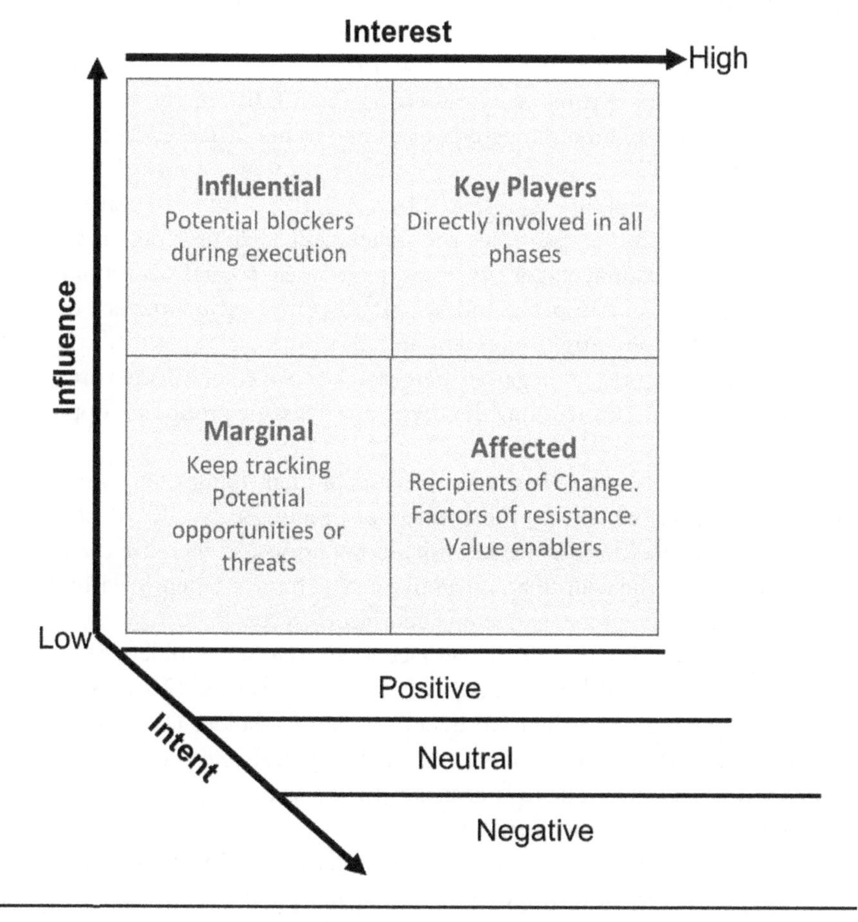

Figure 4.7 Stakeholders Analysis, Influence Grid (Adapted from Johnson, G., & Scholes, K. *Exploring Corporate Strategy* [8th ed.]. © 1997 Prentice Hall)

plans, which will mainly be derived from the positioning of each stakeholder on the influence grid. We don't communicate the same way with a key player, an influential, a marginal, or an affected stakeholder, nor do we communicate the same way with a positive, a neutral, or a negative one. Even within one single group, each specific stakeholder will have its own requirements in terms of communication.

The communication toward the influential stakeholders will be very formal and structured. It is very important to give them exactly what they want, the way they want it. Nothing less, but nothing more either. When performing the analysis, you will see that very often, the influential have a tendency to demonstrate a neutral intent. It comes from the low level of interest. In most stakeholders' management strategies, we'll strive to drive people toward the positive side, but not with the influential. Trying to drag them toward the positive side would require a disproportionate level of effort with a limited result, for a benefit that would be rather small. We will then mostly try to maintain their neutrality, at the same time avoiding letting them fall into the negative side, where they might become dangerous blocking points in the execution of our portfolio components.

That danger potentially represented by a negative influential stakeholder is the perfect illustration of the input the stakeholder analysis represents for risk identification and management processes, be it at the portfolio level or even at the individual project and program levels. If a negative influential is a threat to counter, a positive one might represent an opportunity to exploit. It's the same with the other categories. A negative marginal, key player, or affected are threats to be addressed, and positive stakeholders from the same groups are opportunities to exploit.

One way of addressing these risks consists of identifying their positions on the influence grid, which are the connections among them. We'll be able to know which stakeholders influence which other ones, and then use the positive relays to influence the neutral and negative people more efficiently without having to directly confront a negative preconceived perception.

As described in Figure 4.8, if you face a negative stakeholder (4), but one who has a good relationship with two others who are neutral (2 and 3)—themselves being potentially influenced by a positive stakeholder (1)—you will not directly address the negative one, or even the neutral ones. You will use the influence of the positive one to "permeate" the neutral and propagate a positive influence toward the negative stakeholder.

Finally, I want to make it clear that this classification is not a hierarchy. A key player is not more important than a marginal player, who's not less important than an influential or affected one. In fact, that importance varies with the perspective.

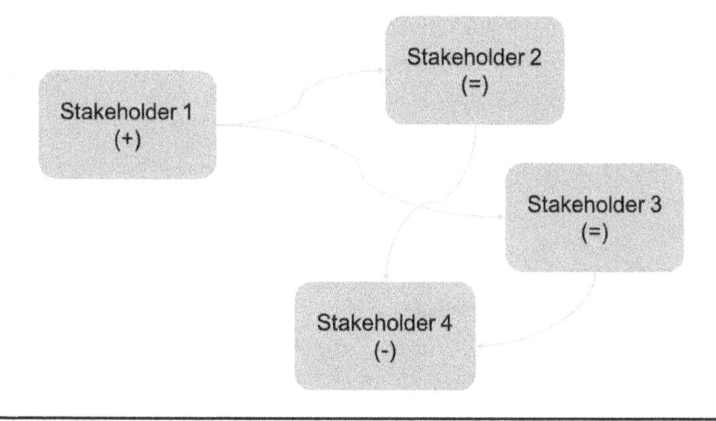

Figure 4.8 Transversal Influence Among Stakeholders

- From a *project* standpoint, the key players will be the most important. The project managers, having to deliver a tangible result, will be mainly and directly working with the key players on a daily basis. They won't be immediately connected to the affected players, who might be out of their perimeter of accountability and responsibility, and won't necessarily be in touch with the influential stakeholders.
- From the perspective of a *program manager*, the most important stakeholders will be the affected ones, as they will be the ones generating potential resistance to change and will be the ones through which the benefits will be effectively realized.
- Finally, from a *portfolio management* perspective, the importance of the influential stakeholders will be higher, because they might constitute blocking points in the generation of the expected performance, and very often, the project and program managers will not hold a sufficient level of authority to address them, relying on the portfolio manager to do so.

4.3.3 Expressing the Expectations

Once we have an exhaustive mapping of our stakeholders, it is time to really talk to them, and more even than talk to them, listen to them.

The idea here is to present them with our initial concept, idea, or vision and support them in expressing their expectations toward it.

This is usually done using brainstorming sessions and techniques. This step of "Expressing the Expectations" involves all stakeholders, wherever they sit on the influence grid. It's important here to avoid any filtering, bias, or judgment with regard to the expectations that will be expressed here. There are no good or bad ideas, nothing possible or impossible. The aim is to collect as many inputs as possible to maximize the number of opportunities to add value to our portfolio and its components as early as possible, before the level of investment reaches a point of no return in the development of our portfolio components and makes it difficult, if not impossible, to integrate a potential value trigger into our scope.

4.3.4 Identifying the Needs

Of course, we will not cover all the expectations having been raised by our stakeholders. First, because some of them will contradict one with another, then because some will be misaligned with our initial intent, and also probably some of these expectations will be unrealistic. We need then to filter the expectation to extract the real potential value triggers, to "Identify the Needs."

What is the difference between a need and an expectation? In fact, not so much. There is no need that has not initially been expressed as an expectation. Needs are then a subset of expectations. But needs represent potential value triggers, meaning something that is (1) potentially feasible, (2) not in contradiction with anything else, and (3) aligned with our initial intent.

To prioritize, or filter, the expectations and identify the needs among them, we will need to use again our stakeholders, but not all of them this time. We'll have to filter our stakeholders according to their intent factor. OK, but which ones will we keep? The idea is to identify potential value triggers.

If we keep the negative ones, their main expectation is, schematically, that you don't launch your initiatives; then they will not provide any additional value indeed. We will use the negative stakeholders later, but not for that part.

If we keep the positive ones, their basic expectations are already fully aligned with our initial intent, and often they are satisfied with what we present; thus, their added value would be quite limited. We will use them in many opportunities, but not here.

The identification of needs will have to be done by using mainly the stakeholders we have initially classified as being neutral. The neutral stakeholders are the ones who will be the most objective, rational, unbiased, and pragmatic. They are the ones who will be able, more than the others, to identify what makes sense and what doesn't, what is the most value-added course of actions.

Using essentially the neutral stakeholders will also be part of a stakeholders' management strategy. There's a psychological bias that makes it very difficult to be opposed to something you have contributed to defining. By using our neutral stakeholders to build what will constitute the heart of our initiatives, we drag them slightly toward the positive intent. This shift will also have an effect on the negative stakeholders. While performing the stakeholders' analysis, you will see that the neutral group is very often the largest one; by involving and even engaging them, it's the biggest crowd that will follow you, isolating the negative stakeholders. Human beings are social animals, we usually don't like isolation. The "rationally" negative stakeholders, seeing the crowd following you, will then have a tendency to apprehend your track as eventually being not so bad; they will then become more neutral, then you'll be able to discuss, engage, and drag them to the positive side. The negatives "negatively negative" will then be totally isolated and, one hopes, at some point be bored of preaching in the desert. It's a marginalization strategy for your more reluctant negative stakeholders and an engagement one for your neutral people.

While identifying the needs, you should be able to refine considerably the number of items in your portfolio cart. If your stakeholders have raised, as an example, 150 expectations, we can imagine you'll end up with a list of 25 needs.

4.3.5 Formalizing the Objectives

Now, with the needs being identified, we have the ideal scenario for our portfolio and our strategy. Of course, the story would be too perfect if we could afford to execute all of the potential initiatives we have identified here, according to our capabilities and our other engagements. As part of the portfolio management exercise, we have to add another layer of prioritization, deciding which of the needs we have identified will actually be converted into concrete and tangible action items during our next strategic cycle, be it for the next time period to come or even up to the organization's strategic horizon. In other words, we need to formalize the objectives among the needs we have identified. What's the difference between a need and an objective? Not so much, in fact. There is no objective formalized that has not been previously identified as a need; objectives are then a subset of the needs, but an objective is a need which the organization decides to implement and commits to do so in front of the stakeholders. A need is, "We *could* do this"; an objective is, "We *will* do it."

To prioritize, we enter into the very mechanism of portfolio management, into the core of this pillar of strategic alignment. We will have to identify which elements, or combination of needs, being satisfied will generate the highest value, which will contribute the most to the achievement of the organization's strategic vision, and which will serve the most the very strategy of the portfolio itself.

The first step to this prioritization in regard to strategic alignment, is of course, to have a strategic vision, goal, or objective properly expressed in qualitative terms, as we have explored in the previous chapters.

Depending on the level at which we place our portfolio, this strategic expression can illustrate different depths in the very strategy of the organization.

If we are at the highest strategic level of the organization, the enterprise level, we address here the rationale for the existence of the organization; we're then at the very top level of the construction of this opportunity chain, leaving only one level of alignment, but making it more complex to express. This top enterprise level constitutes a portfolio—often a portfolio of portfolios and major programs. Within each of these sub-portfolios and programs, we'll have to decide which of their respective components will serve the global strategic purpose and the specific or localized strategy of the sub-portfolio or program. The construction of that subsidiary opportunity chain will then start from there, connecting each portfolio component to the upper level of strategy.

4.3.6 Determining the Strategic Contribution of Each Candidate Component

A strategy is expressed through the formulation of a strategic vision, translated into tangible capabilities to develop within the organization to realize that vision, the necessary means to put it in place, allowing the organization to shape its business environment and develop a unique competitive advantage. Every initiative undertaken within the organization should then contribute to these strategic objectives, to these, as we name them, "business drivers." Based on the relative prioritization of these business drivers among themselves, which can be determined using a paired-comparison matrix, we'll determine the most suitable mix of needs to be formalized as objectives and thus determine, at a later stage, the corresponding components to be included in our portfolio.

Business Drivers or CSFs	A	B	C	D	E	Total	Weight
A: Improve client retention	X	4	5	5	5	19	40%
B: Deliver projects	1	X	4	5	5	15	30%
C: Improve ROI	0	1	X	4	3	8	15%
D: Improve Employee Satisfaction	0	0	1	X	4	5	10%
E: Secure Compliance	0	0	2	1	X	3	5%

Figure 4.9 Paired Comparison Matrix

The paired comparison matrix is easy to understand and apply: If we want to compare a set of factors among themselves, we'll simply constitute each possible pair of factors and distribute 5 points among each of them. While comparing A with B, if A is much more important than B, we'll give 4 points to A, leaving 1 point to B—4 plus 1 equals 5. We'll then sum the scores and extract the corresponding percentages. These weightings will allow us to establish a baseline representing the total value or benefits to be created at the specific organizational level. If the portfolio we're addressing is positioned at a lower level of the organizational framework, such as a departmental or divisional portfolio or a product portfolio, we'll have to prioritize two levels of critical success factors: the organization's overall business drivers and the portfolio-specific critical success factors to which each of our components will have to contribute. An example of a paired comparison matrix is illustrated in Figure 4.9.

4.3.7 Constructing the Strategic Alignment Model

Once we have the relative priorities among the business drivers to which our portfolio components should contribute, we'll be able to identify and quantify that contribution in order to define which needs or combination of needs will serve the strategy best.

That assessment of strategic alignment consists of giving a grade to each need (or directly potential component) according to what we assume as being its direct contribution to each and every single business driver. That grade can be defined on a scale from 0 to 10, 0 meaning no contribution at all, and 10 meaning an essential contribution.

It is important to assess and grade each need toward each business driver, as we're looking for the best portfolio mix to maximize the achievement of our strategic objectives and thus the creation for our organization and its stakeholders.

Once we have scored our different identified needs, we multiply the grade given to each by the weighting of each corresponding business driver, and we obtain an individual score for each identified need. When summing the scores obtained for each of them, we'll have a global scoring for each identified need, allowing us to see which one or which ones should be formalized as objectives—meaning what will drive and shape the mix of our portfolio, at least for the next cycle to come.

This is because formalizing a current set of identified needs as objectives, for example 10 objectives among 25 needs, doesn't mean the 15 remaining needs will never be covered. If these stakeholder expectations have been identified as potential needs, it means they present a certain value for the organization; it would then generate an opportunity cost in not covering them at some point in

Weight / Scenario	Business Driver A	Business Driver B	Business Driver C	Business Driver D	Business Driver E	Total
Weight	40	30	15	10	5	100
Scenario 1	3 / 120	4 / 120	8 / 120	3 / 30	6 / 30	420
Scenario 2	7 / 280	4 / 120	2 / 30	5 / 50	8 / 40	520
Scenario 3	6 / 240	7 / 210	3 / 45	5 / 50	1 / 5	550
Scenario 4	4 / 160	4 / 120	3 / 45	9 / 90	2 / 10	425
Scenario 5	2 / 80	7 / 210	4 / 60	6 / 60	9 / 45	455

Figure 4.10 Strategic Alignment Assessment Matrix

time. We still keep the opportunity opened to integrate these other needs into our "delivery" framework at some point in time—(1) when enough of the most valuable items will have been covered, (2) if we act upon them to maximize their strategic contribution, or (3) if any additional resource would be obtained within the organization allowing us to expand the scope of our portfolios. Figure 4.10 shows an example of a strategic alignment matrix.

In the example in Figure 4.10, we can see that scenario/need 4 is the one that will definitely have to be formalized as an objective, as it represents the highest contribution to the business drivers of the organization. It's followed by need 2, and further the other ones.

We know now which elements we'll have to cover within the content of our portfolio within our strategic horizon, what will be the basic needs of our stakeholders which will allow us to generate the highest value and create most of the benefits we can expect to create. In other words, we can claim what are our objectives. But at this stage, their formulation may still be unprecise and quite high level. In order to structure our portfolio and translate these objectives into elements which will be quantifiable and measureable and will deliver tangible results, we need to push the level of detail of the opportunity chain a little bit further.

4.3.8 Functional Analysis, Critical Success Factors, and Key Performance Indicators

With the stage of "formalizing the objectives" in the construction of the opportunity chain, we have defined what would correspond to the scope—the perimeter of our portfolio, but not yet to its very content. Each of the objectives we have just formalized is still a qualitative statement, sometimes generic and general. Now is the time to come back to our stakeholders, to the ones that have expressed initially the expectations from which we have derived our formalized objectives. To clarify these objectives, we will ask our stakeholders to play a grammatical game, which most IT people know well: *functional analysis.*

What is a function, grammatically speaking? It consists of an active verb with a measurable noun. "Sustaining the weight" (of the person sitting on it) is the typical example of one function of a chair. Anything can be translated into functions. In Agile project management methodologies, these functions are often called *user stories,* which represents quite well the underlying idea behind them. Each objective will be translated, worded, as a function. It's a one-to-one relationship at least, if not one to many. Depending on the complexity of the elements, we might need to use several functions, or even sub-functions, to translate one objective. That functional analysis can be graphically represented using a *functional breakdown structure* (FBS), as illustrated in Figure 4.11.

This functional analysis allows us to align everyone concerned on the same understanding and perception of the objectives to reach, and how reaching these objectives will have to be materialized and made concrete once the results delivering these functions are delivered by the various related portfolio components.

Now, having functions defined is a good thing—it clarifies and starts to make things tangible. We have already moved from the fuzziness and high ambiguity of our initial concept or strategic vision to something which starts to be tangible, looking concrete and manageable. We need to make sure that we can use these developments to quantify and not only qualify the value to be created. The last two elements to consider achieving that "measurability" are part of a prioritization exercise. Exactly as business drivers have been determined at the strategic level of the organization, we need to identify the specific critical success factors for our portfolio—the most important elements of our strategy to which any component of our portfolio will have to contribute. A critical success factor (CSF) designates a function which is more important than the other ones,

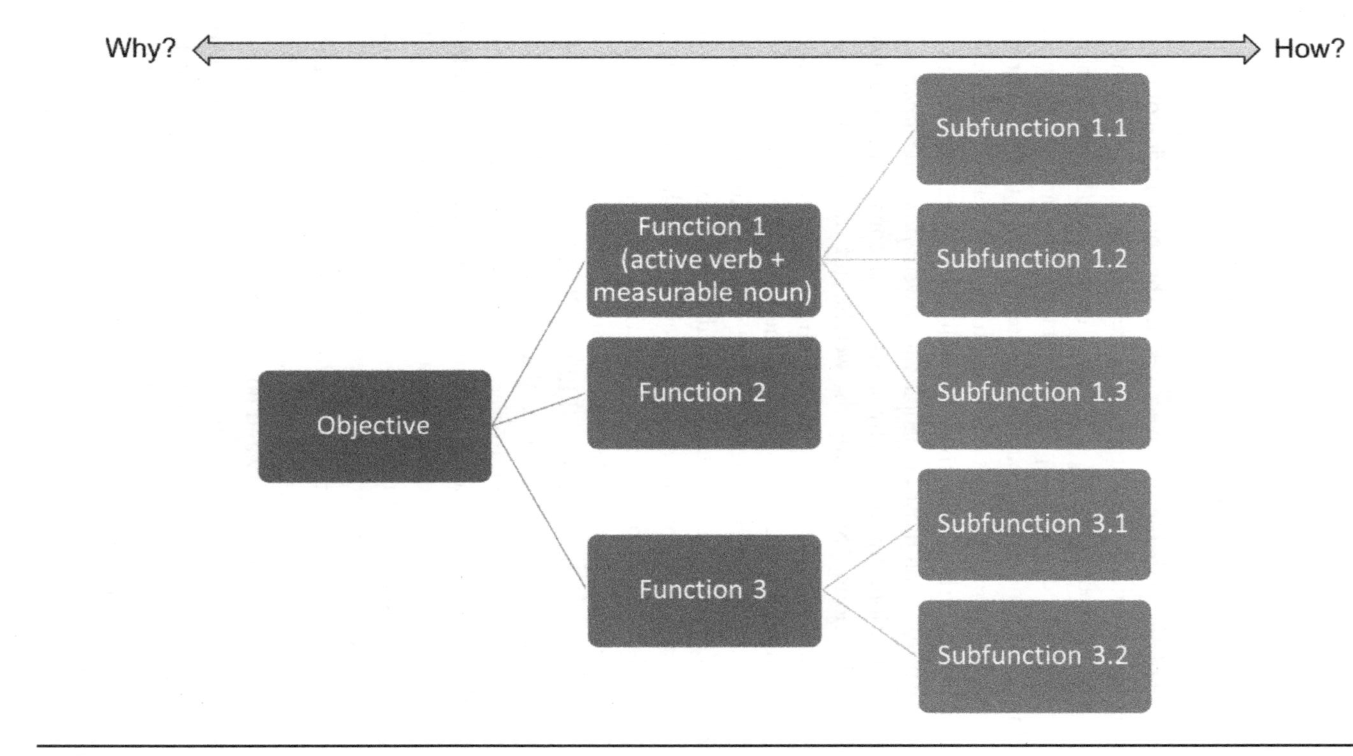

Figure 4.11 Functional Breakdown Structure (FBS)

in the sense that this function is essential to the whole. Without these functions, the other ones identified become useless. A CSF is a characteristic of a function, not a subset. A function is or is not a critical success factor. There will always be at least one critical success factor, more often several of them; but all of the functions can't be critical. If everything is critical, then nothing is.

These critical success factors will constitute our "qualitative" quality indicators.

To create value, we'll have to at least deliver the critical success factors within the timeframe of our organizational strategic horizon. An item which will definitely not be delivered within the strategic horizon shouldn't be a critical success factor. It shouldn't even necessarily have been formalized as an objective. But as we move forward within the cycles of our portfolio management exercise, pushing our strategic horizon by one year every year, needs which have not been formalized as objectives might be formalized when we review the opportunity chain, and objectives or functions which have not been considered as CSFs at some point in time could change status. In one way or the other, an element which was considered as a CSF within one cycle of the portfolio management exercise could be considered as not critical during the next cycle, necessitating review of the priorities among the different components addressing the various functions described in the opportunity chain.

But any function, active verb and measurable noun, critical success factor or not, is a binary indicator, zero or one. Binary does not represent a measurable value as such in the sense we described it earlier. Then we need to define a key performance indicator (KPI) for that function. A KPI will be constructed around a criterion, which represents what we effectively measure. That criterion has to reach a certain level, within a range of flexibility. If you're designing a new robot vacuum cleaner, one of the functions will consist in "covering a sufficient area on one battery charge." That function will probably constitute a critical success factor, and one of its KPIs could be

Criterion: surface to cover in square meters
Level to achieve: 150 (m^2)
Flexibility: ± 5 (m^2)

We know now that our new robot vacuum cleaner has to cover an area between 145 and 155 m^2 to deliver the expected and acceptable performance. If when testing it, we see it covers only 140 m^2, then we know it is performing below our quality and performance target. If the vacuum cleaner dies after having covered 160 m^2, we know we have probably used a battery that is overperforming and that probably cost more than it should have—we've done some "gold plating" and jeopardized the value balance.

4.4 Integrating the Opportunity Chain throughout the Organization

The aim of the opportunity chain is to move from a relatively fuzzy, blurry, undefined initial concept and refine it into tangible and measurable items (see Figure 4.12). Tangible because each function defined in the opportunity chain will have to be covered by one (or more) component of our portfolio, allowing us to construct the mix of our portfolio. Also, it will lead us to the definition of each component itself, the objectives of those components being to fulfill the corresponding function(s) and to do so according to the performance documented within the related KPI(s). The establishment of critical success factors gives us a first layer or prioritization highlighting the most important elements and clarifying what constitutes the value to be generated at that level. The opportunity chain is thus a tool allowing us to reduce ambiguity and decrease the level of complexity in the elaboration of the strategy. But it's also the main hinge to connect the different layers of strategy and portfolios, programs, and projects.

In any organization, at the upper governance level everything comes (or is supposed to come) from the expression of the strategic vision. Realizing this strategic vision involves and impacts a certain set of stakeholders—at this level designated as business environment stakeholders—which we have to identify and somehow, at a high level, to categorize using the 4i's model. These stakeholders will have a certain number of expectations with regard to the particular business area addressed. Following the expression of these expectations, we will identify market needs, as portions of the market that might effectively represent a potential value and are in alignment with our initial strategic vision. To decide which of these objectives will be effectively integrated as part of the strategy, we need to rely on the assessment of our organizational capabilities, on the analysis of the current initiatives and investments, and on the evaluation of the overall current performance of the organization. Covering these aspects and providing the inputs for the decision-making process, which will consist of formalizing the strategic objectives of the organization, is nothing less than the very aim of the portfolio management exercise. This is mainly how we formalize the strategic objectives. The following step consists in clarifying these strategic objectives into statements describing the organizational capabilities to construct (meaning to build, such as in industrial capabilities, new products or services or new markets to open) and/or the business benefits to develop (enhancing project management abilities, developing quality compliance policies, optimizing the organizational structure, etc.). These strategic objectives can of course be classified and prioritized by identifying which among them represent strategic critical success factors, also called *business drivers*. We'll determine key business performance indicators corresponding to these business drivers, defined

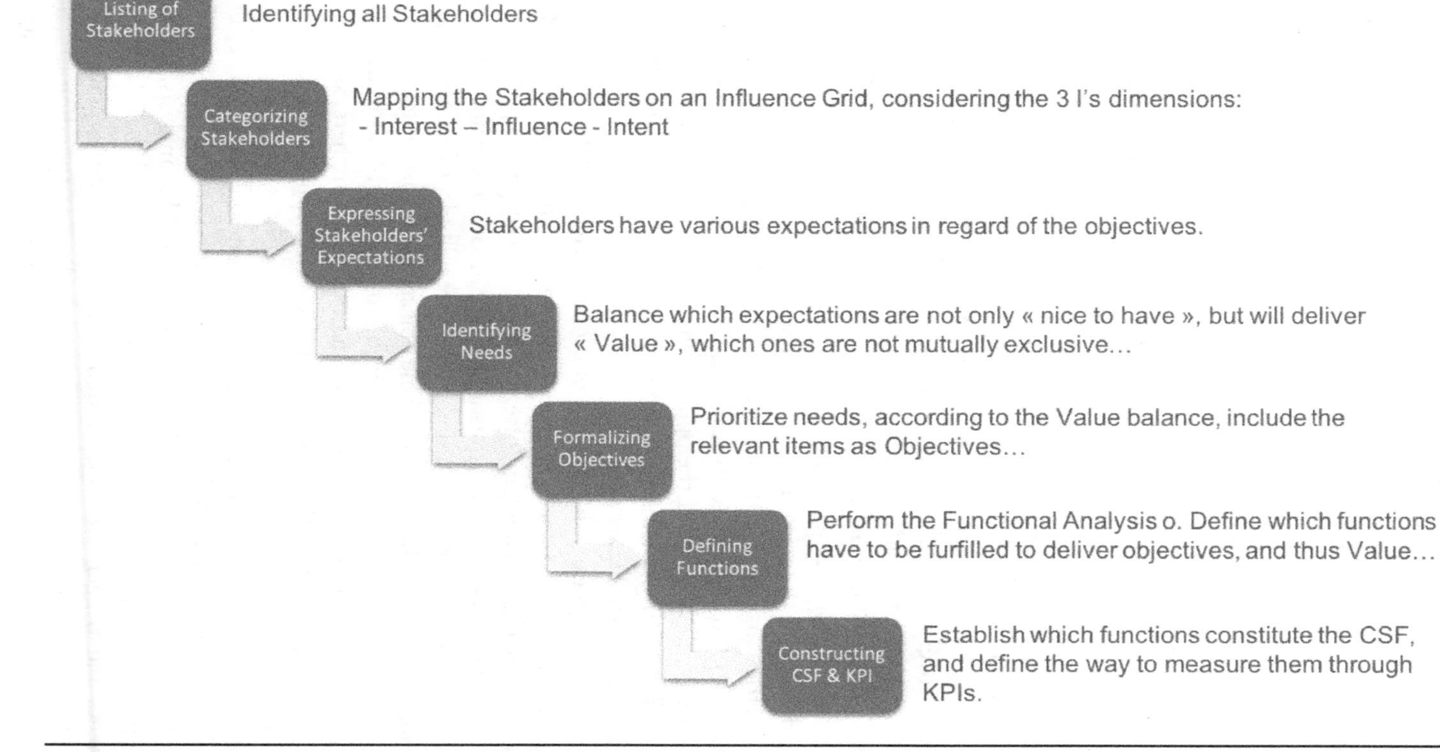

Figure 4.12 Development of the Opportunity Chain (Reproduced from Lazar, O. When Change Is Not a Change Anymore: Organizational Evolution and Improvement through Stability. PMI® Global Congress 2016—EMEA, Barcelona, Spain. Newton Square, PA, USA: Project Management Institute. © 2016 Olivier Lazar)

around a criterion, having to achieve a certain level within a determined flexibility. What we have just described here is nothing less than the application of the opportunity chain to describe the organization's strategy, describing the set of initiatives to undertake to realize the organization's strategic vision (see Figure 4.13).

Delivering each of the formalized strategic objectives, translated as organizational capabilities or business benefits, implies launching one or more corresponding initiatives. These initiatives will need to take into account a certain number of stakeholders (often a subset of the business environment stakeholders), who need to be categorized using the 4i's model. These stakeholders will have a certain number of expectations with regard to the development of the corresponding strategic objective. Among these expectations we'll identify some needs that represent potential triggers to enhance that strategic objective. According to the capability and resources allocated within the portfolio management exercise, we will filter these needs in order to formalize objectives to cover within the current initiative to develop the organizational capabilities or realize the business benefits corresponding to the strategic objective. By stating the functional characteristics of each objective, we will then be able to identify which ones will represent critical success factors for the current initiative, determine the related key performance indicators, and derive from that the specific tangible results supposed not only to be produced within this initiative but also to contribute to realizing the strategic objective. Again, we have developed a lower level of the opportunity chain for each strategic objective, describing nothing less than the different programs having to realize these strategic objectives (see Figure 4.14).

Within these programs, the delivery of each required specific and tangible result expected to be the benefit triggers will again mean taking into account a certain number of stakeholders (often a subset of the program-level stakeholders) who need to be categorized using the 4i's model, and who will express a certain number of expectations, among which we will identify a certain set of needs, among which we will formalize some objectives. Translating these objectives into functions (active verb + measurable noun) will enable us to determine critical success factors and key performance indicators. We are here, as you have certainly already guessed, at the project level. Here, the opportunity chain will allow us to literally shape the project (see Figure 4.15).

The core elements of the opportunity chain, from the expected result as stated at the program level with the corresponding function and the related KPI down to the formalized objectives through the stakeholder analysis, will feed the initiation of the project. The functional analysis will feed the elicitation of requirements and trigger the construction of the project scope statement by defining the concrete deliverables having to cover the corresponding function. Critical success factors and key performance indicators will fill in the quality

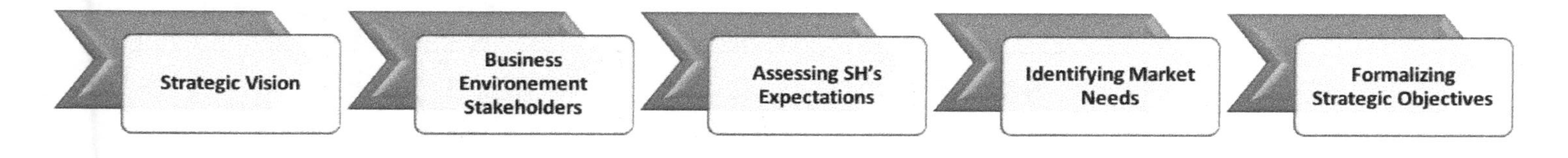

Figure 4.13 Strategic-Level Opportunity Chain

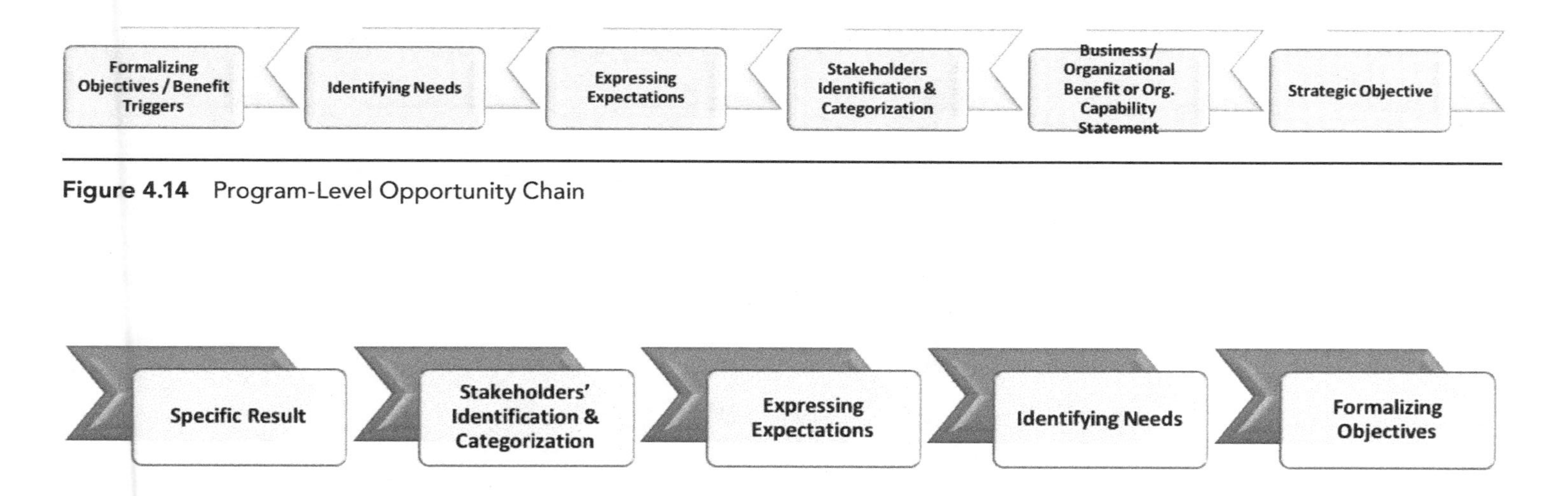

Figure 4.14 Program-Level Opportunity Chain

Figure 4.15 Project-Level Opportunity Chain

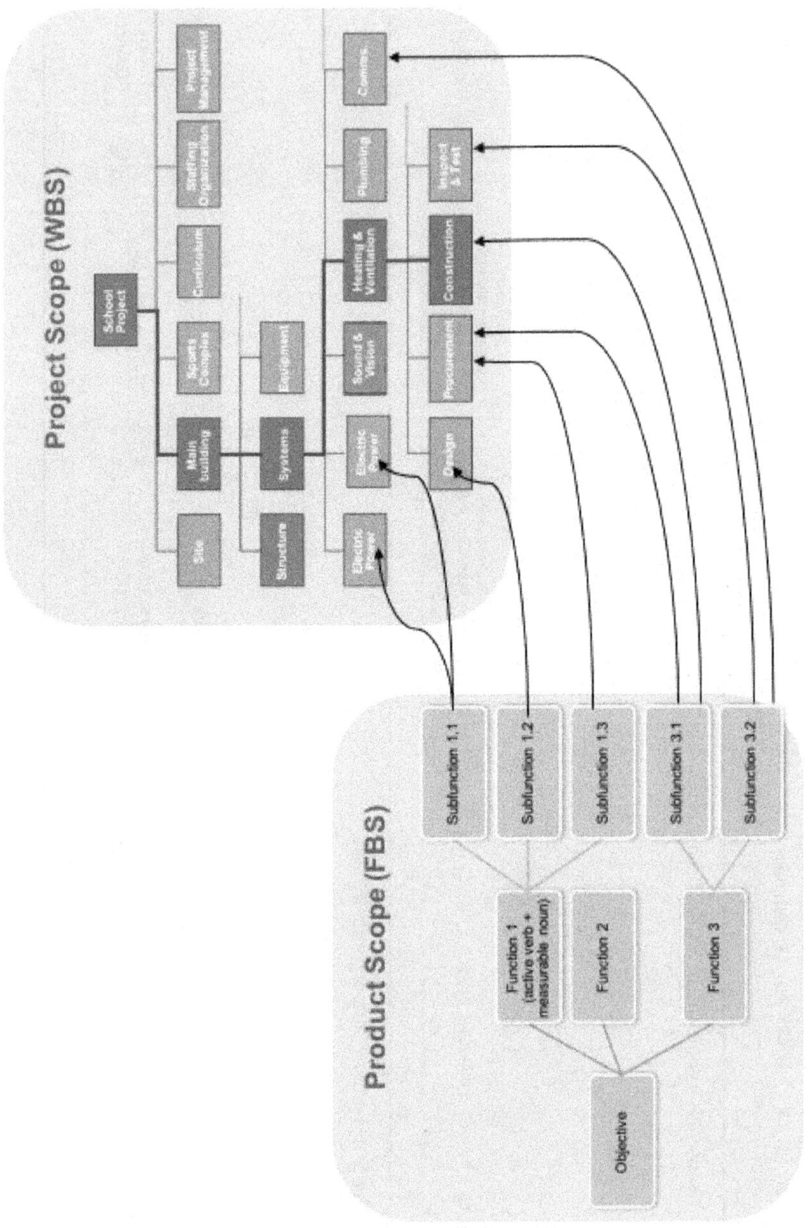

Figure 4.16 Product Scope and Project Scope

(See discussion regarding Figure 4.16 on page 94.)

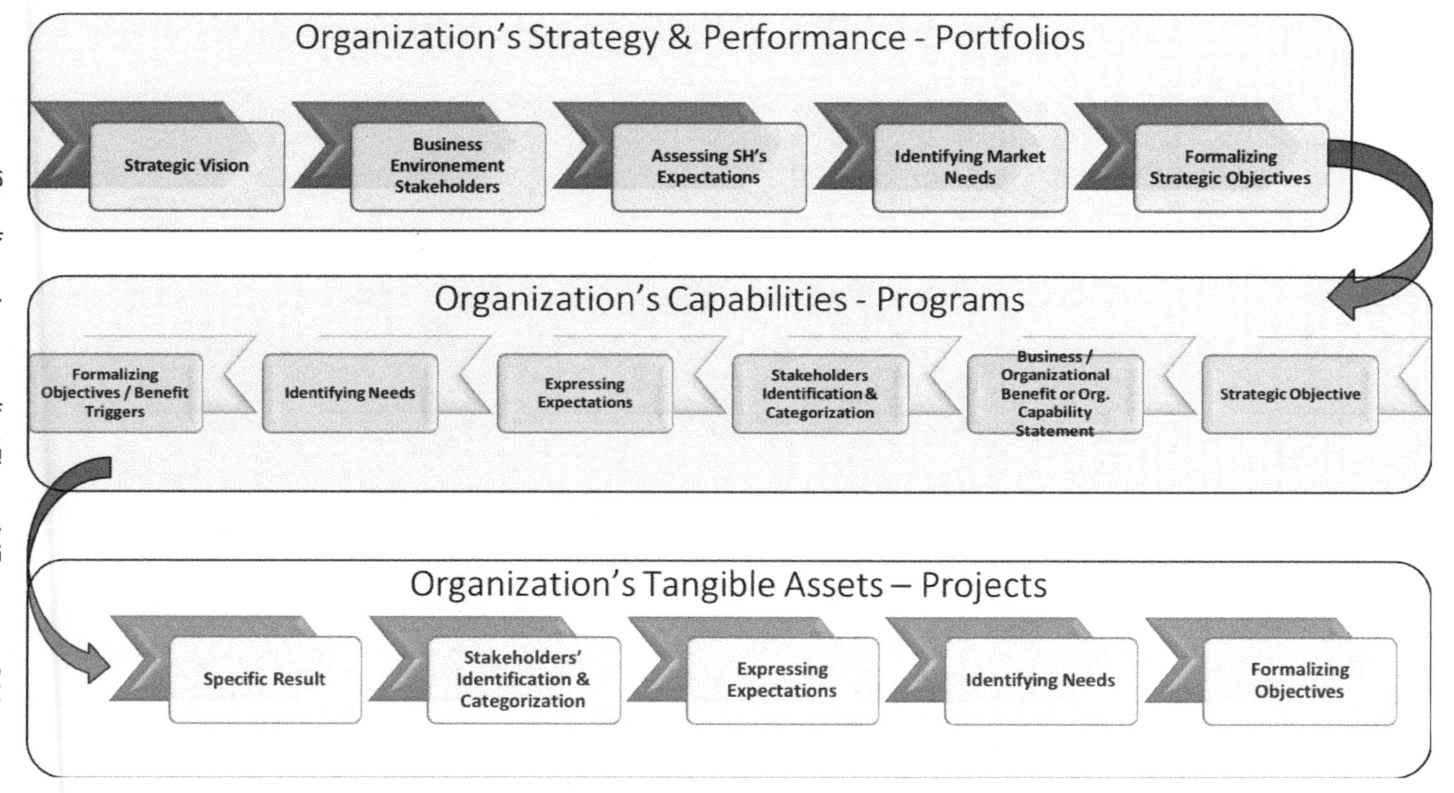

Figure 4.17 Integrated Opportunity Chain throughout the Whole Organization

(See discussion regarding Figure 4.17 on page 94.)

checklists, the CSFs being the "qualitative" value realization indicators and the KPIs being the "quantitative" quality metrics. These tangible outcomes to be produced within the projects will be delivered by work packages, detailed in the project's work breakdown structure (WBS). At that level, the functional breakdown structure represents the *product* scope and the WBS represents the *project's* scope (see Figure 4.16).

At each level—strategic, program, or project—the stakeholder analysis will also feed the communication plans for each initiative and somehow contribute to the overall risk identification.

The opportunity chain can be deployed throughout all governance layers within the organization (as seen in Figure 4.17) and allow us to ensure and secure the strategic alignment of each component of the portfolio with the strategy. It even represents the strategy itself and is to be used to communicate that strategy to the stakeholders, including clients, employees, and shareholders; and of course managers and decision makers; portfolio, program, and project managers; and PMO leaders who will have to deliver the strategy, realize the strategic vision, and sustain the organization.

Chapter 5

The Third Pillar: Risk

As essential as strategy could be to determining a direction to lead the organization, and as key the strategic alignment factor in the selection and prioritization of portfolio components can be, an additional pillar is necessary to support the construction of that portfolio.

That pillar is *risk*. Every single decision we make is based on a risk analysis. Anything we do, from the smallest movement to the most important life decision, is made either to counter a threat or to exploit an opportunity. Even reading this book is a risk-based decision you've made. Either you read it because you're expecting to find answers to a current portfolio management–related problem you're encountering (that's threat mitigation), or maybe you're just curious and expect to learn something new that might be useful (that's enhancing an opportunity).

At the portfolio level within a governance framework—the aim being to operate business and organizational capabilities to generate the highest level of performance—we also have to manage the exposure of the organization to risk. We have to balance between risky and rewarding investments and secure, but maybe less rewarding, ones.

One of my most provocative statements is to say that basically, if you don't manage your risks, you don't manage anything. Often risk is the last thing on the list when addressing projects, programs, and portfolios. But risk management is what makes the difference between success and failure. Often the deviations noted on projects and programs are nothing less than risks which have not been identified, or often, which is worse, intentionally ignored.

But what do we mean by risk? There are many definitions. When we type "definition of risk" in Google, there are 31,200,000 results—probably not as many different definitions, but we can at least expect to have a few hundred of them.

When speaking of project, program, and portfolio management, the most commonly admitted definition is the one formulated in the *PMBOK® Guide*: "Individual Project Risk is an uncertain event or condition that, if it occurs, has a positive or a negative effect on at least one project objective.

"Overall Project Risk is the effect of uncertainty on the project as a whole, arising from all sources of uncertainty including individual risks, representing the exposure of stakeholders to the implications of variations in project outcome, both positive and negative" (PMI, 2017a).

Risk, in common language, even in business language, is often perceived as a negative element, a danger, a probability to harm or being harmed. It's a natural trend. The human brain is wired to see the threat first, since the times when our ancestors were hunting in the savanna, mainly guarding against being eaten themselves by a saber-tooth tiger. Perceiving the negative side first is one of those favorable evolutionary factors that allowed us to survive until now. But what is true to survive in the savanna can actually be counterproductive in the business world. Even if identifying and countering, as much as possible, threats putting our businesses in jeopardy is important, we'll see that looking for opportunities and ensuring we do everything we can and even focusing our resources on exploiting these opportunities is more important. Value hides in opportunities, and instead of calling that "risk management," calling it "opportunity management" would definitely make more sense here. As Peter Drucker said, "Effective strategies should be focused on maximizing opportunities, and action should not be based on minimizing risks, which are merely limitations to action" (Thiry, 2004). Then it's not only a matter of definition, of processes, of governance, it's also a matter of mindset. And that mindset is not easy to change, driven by seven million years of gradual evolution.

Let's see how it works.

5.1 Specificities of Risk Management at the Portfolio Level

The exposure to risk at the portfolio level is mainly the result of the combination resulting from the consolidation of the various risks which have been identified at the component level and the risks specifically identified at the portfolio level, often then related to the realization of the organization's strategy and generation of the expected performance. The consolidated risks we will call the *portfolio intrinsic risks*. The strategic risks being mostly triggered by the business environment, we will call them here *portfolio extrinsic risks*.

5.1.1 Specificities in Risk Identification

The identification procedure of risks will then focus on different aspects. The intrinsic risks could come from a variety of sources:

- **Portfolio risks.** These are the risks identified at the portfolio level, triggered by specific portfolio activities, such as the transitioning of capabilities from the programs that put them in place into the operations that will exploit such capabilities, or risks related directly to the portfolio specific decision-making process.
- **Aggregated risks.** These are the risks impacting more than one component of the portfolio, even if triggered by a single component. These risks are then better managed or at least controlled at the portfolio level, even if the treatment activities of these risks can be executed at the level of the components triggering the risks.
- **Component risks.** These are the risks impacting a single component, but if the impact of those risks exceeds the perimeter of accountability of the component manager, then this accountability and the related decision-making is escalated to the portfolio level.

Any of these risks can also be classified as:

- **Operational risks.** Risks triggered by the execution of the portfolio activities or components. The inputs supporting the identification of operational risks are mostly related to the execution of the portfolio components, then we will use tangible and mostly quantitative elements:
    - Work breakdown structures of the projects and the mix of program components.
    - We'll look at the inner parameters and constraints of each component related to the quantities of time and resources, the critical paths and their variations, etc.
    - Shared resources among components in the portfolio can also create external dependencies among these components, which become factors of risk.
    - Indeed, historical data and their availability is critical in this context. No component manager should start a risk identification workshop without having in hand the risk registers of other previous and similar

projects or programs; even the risk registers from the other components of the portfolio can also be useful.
   - The estimates related to any quantity related to the components have also to be considered. An estimate is always linked with a certain accuracy or uncertainty. Then that level of accuracy represents the level of risk taken or tolerated on that estimate. If you make an estimate whose targeted accuracy is 5 percent, it means you tolerate 10 percent of risk on that estimate.
   - Your portfolio procurement plan is also an input to the risk identification. Subcontracting one component or part of a component may be a response to a risk, but it will also trigger the appearance of other risks.
   - And last but not least of the operational risk identification inputs: the change requests processed at any level within your portfolio. Anything you change within your portfolio will be a factor of risk. It may eliminate some risks, absorb the impact of some others, but will also trigger new ones. And change is permanent in a portfolio, in a program, or in a project. Change is the inherent constituent of project management, as is risk management.
- **Contextual risks.** Risks related to the environment of the components. Contextual risks can also be extrinsic, but intrinsic contextual risks are the ones triggered by the environment created by the context of the portfolio itself around its components. They can be considered extrinsic from the perspective of the components, but intrinsic from the perspective of the portfolio. The inputs supporting the identification process for contextual risks, intrinsic or extrinsic (see Figure 5.1), are mostly based on environmental elements:
   - The stakeholder analysis and the 4i's grid, at the component and portfolio levels, will give us an interesting set of information. Overseen from the risk perspective, a negatively affected stakeholder is a threat (negative risk) which can generate resistance to change, putting in jeopardy the realization of benefits, and thus the generation of the expected business performance. This is the same with a negative influential stakeholder. These threats need to be addressed, as a positive and marginal stakeholder could represent an opportunity which would be wise and interesting to exploit.
   - The variation in the portfolio critical success factors, their levels of priority, and the similar changes in the organizational business drivers will have an impact on the prioritization of portfolio components. The level of the business environment evolution rate, used to calculate the strategic horizon, is important to consider. The higher it is, the higher is the level of risk in changes in that environment, the less stable will be the business drivers, and thus the portfolio critical success factors.

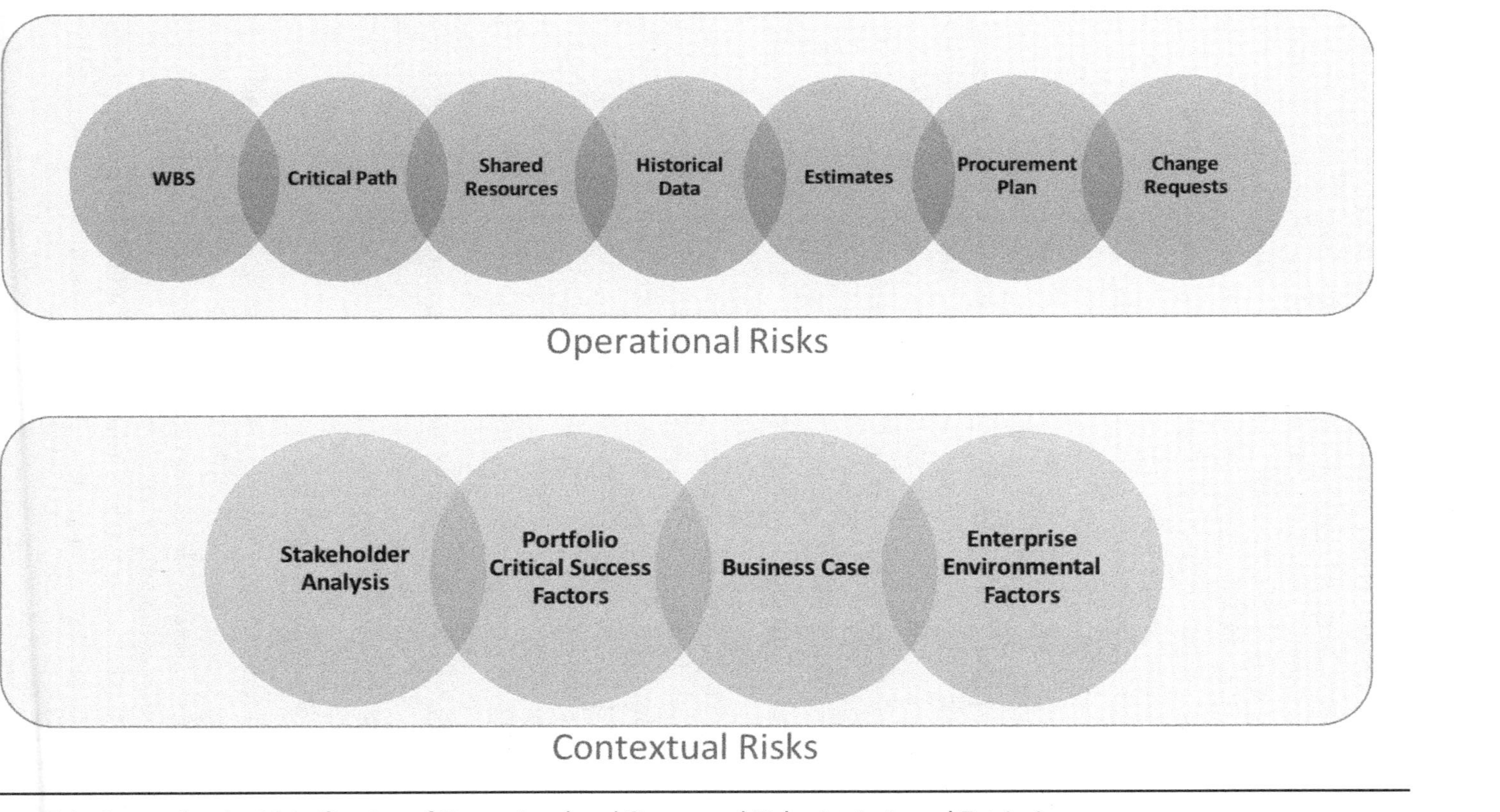

Figure 5.1 Inputs for the Identification of Operational and Contextual Risks, Intrinsic and Extrinsic

- ○ The inherent uncertainty related to the performance objectives of the portfolio and its components, as described in the overall and specific business cases, will also have to be considered as a factor of risk.
- ○ And finally, the so-called Enterprise Environmental Factors, as described in the *PMBOK® Guide* (PMI, 2017a), including the regulatory environment, the organization's strategy, even the very organizational structure and culture of that organization—in other words, everything shaping the environment of the portfolio and its components—can be perceived as a factor of risk. This particular aspect most often leads to the identification of extrinsic contextual risks.

A common tool used to feed the risk identification process consists of the SWOT (Strengths/Weaknesses/Opportunities/Threats) analysis to assess the organization's readiness in implementing the different components of its strategy and portfolios. Even if that tool is commonly used and seems to be well known and advertised, it is still worth highlighting some important aspects of it.

5.1.2 SWOT Analysis Explained

Addressing SWOT analysis here can seem like a basic exercise, as everyone should be quite familiar with this concept, which seems to be itself relatively simple and straightforward. But I have seen this basic tool being used and misused often enough to consider spending some time in digging into it.

As mentioned above, SWOT stands for Strengths, Weaknesses, Opportunities, and Threats. It's mainly aimed at supporting the identification of risks that could impact the realization of the strategy and the deployment of portfolios. But in fact, it's much more than that, and it can be a very powerful tool to develop a whole risk management strategy for our portfolios, even at the enterprise level.

Let's look at the identification part first:

Strengths. What can we identify as strengths?

- Our various existing capabilities, such as products and services; and organizational capabilities, such as the existence and availability of a PMO; or formal competency development strategies; or an organizational culture and structure supporting our aims.
- Our tangible assets, such as equipment, buildings, materials, financial reserves, intellectual property and patents, brand reputation, global presence, etc.
- Our competitive differentiators.

- Of course, and obviously, people. Yes, people are an organization's major strength. Their motivation, their engagement are an organization's most precious assets and competitive advantage.

Weaknesses. What would constitute weaknesses for an organization?

- Can be financial: lack of funds, low margins, high debt, etc.
- Can be organizational: inappropriate organizational structure, dispersed teams, resistance to change, etc.

Opportunities. Which opportunities can we identify?

- Opportunities related to the market's level of maturity.
- Opportunities related to technology.
- Opportunities related to the organizational learning and self-improvement processes.
- Opportunities related to research and development initiatives, innovativeness.
- Opportunities related to business development.
- Opportunities related to digitalization, globalization.

Threats. Where might a threat for the organization eventually come from?

- Can come from the environment (unstable regulatory or business environment, short-termed strategic horizon; level of competitiveness on the market; legal or regulatory obstacles, etc.).
- Can come from obstacles (regulatory or business, or technological).
- Can come the operational side of our initiatives.

Identifying these elements is not that difficult, as the SWOT analysis is a common tool used in the elaboration of any kind of strategy (see Figure 5.2). But often, this is where people stop the exercise. At the identification level. There is much more to get from a SWOT if we push the exercise a bit further. The aim of SWOT analysis is, in fact, to develop plans—plans that will pursue the following objectives:

First, making sure that our strengths match with our opportunities and allow us to exploit these opportunities by using our strengths by developing the appropriate set of actions to extract all the potential benefits identified within the opportunities. This can, and probably will, lead us to identify new components within our portfolio and will certainly have an impact on our business cases.

Second, these plans will have to trigger some transformations of weaknesses into strengths by developing the necessary course of action to make sure we're addressing these weaknesses to avoid being confronted with blocking points related to these weaknesses. They will also trigger some transformation of

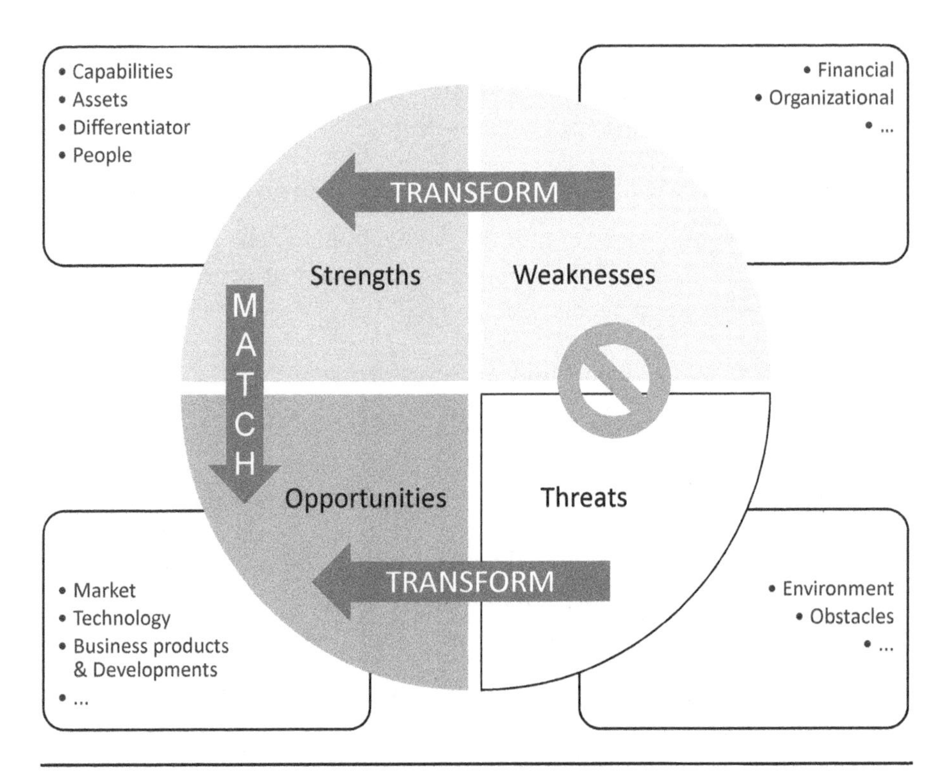

Figure 5.2 Principles of SWOT Analysis

threats into opportunities by considering these elements from the perspective of the benefits to be obtained from confronting these threats and developing the appropriate risk-management strategy.

Finally, we'll make sure that our plans derived from the SWOT analysis allow us to isolate threats from weaknesses, ensuring that the threats don't become weaknesses that might become blocking points, and weaknesses don't become threats for which we'll have to consume a set of precious resources that will unbalance the value equation.

One last important aspect to understand about risk identification is related to the connection between risks and the complexity of the portfolio components, and more precisely with the level of ambiguity of these components.

We have detailed in earlier sections of this book that complexity is the combination of uncertainty and ambiguity—uncertainty being a lack of quantitative information (mainly related to time, cost, resources, effort, etc.), and ambiguity being the lack of qualitative information (mainly related to the definition of the end result or outcome of a certain component or the process to obtain that outcome or result). It's an obvious statement, as its very definition indicates that it is an uncertain event or condition, that risk is related to uncertainty. It's triggered by uncertainty (e.g., level of accuracy of estimates), and it also triggers uncertainty at the level of the component by adding the individual levels of uncertainty of each specific risk (their probability) to the overall uncertainty of the whole component. It's a kind of self-feeding factor.

But risk is also related to ambiguity. Risk identification requires a certain number of data, if not information, to be effective and as exhaustive as possible. Of course, exhaustiveness in risk identification does not exist—you will never be able to identify all of the risks in a particular component of your portfolio; that would require mediumistic abilities, and we're talking here about management science, not science fiction. Then the more ambiguous is a certain component in your portfolio, the less available information you have to identify risks, the less accurate will be your risk analysis, and the less efficient will be your risk management endeavor. If you're less capable of identifying risk, it doesn't mean that your component is "less risky," it just means that it's more exposed to the famous "unknown-unknown" events, which you are unable to anticipate, but which will still happen, directly in connection with the well-known Murphy's Law.

The statement describing the relationship between complexity and risk management is then quite simple: the more ambiguous is a component of your portfolio, the less capable you are of identifying risk. The less ambiguous is that component, the more information you have available to identify risks, and the more you introduce uncertainty within that component.

It then becomes very important to be able to equip ourselves with the relevant risk management structures to face these factors. And that starts with risk identification.

5.1.3 Defining the Appropriate Risk Management Governance Structure

First, while using the different inputs and risk categorization we just have described in the section above, it's important to differentiate the kinds of events impacting the execution of a portfolio component, project, or program. Knowing what is a risk, and what is not a risk, enables us to differentiate between the risk and its consequence. In each of my workshops and seminars on risk management, I always ask this question: "Do you have a risk of being late in the morning?" Besides making the ones who effectively arrived late at the seminar blush, the reactions are quite unanimous: People agree that they risk arriving late in the morning. Here is the trap: there's no risk of arriving late. Arriving late is the consequence of what made you be late. There's nothing you can do about being late, once you're late, you're late, period. But you could have had an impact on the factor that made you be late. And that factor was the real risk—that thing you could have influenced.

A second common flaw consists in confusing risk and defect. When in the risk register of an IT system development project, I've seen an item saying: "Risk of bugs," which made me smile. A bug in software development is not a risk, it's a defect; you're supposed to do your development job properly. Along the same lines, a common risk found in the registers (when they exist) is "Risk of delivering late/above budget." No! These are also defects. Defects in your process, but still defects. Not risks. A last funny one, "Risk of lack in resources." Still a defect in your planning, and worse, in your portfolio management process. You are supposed to plan your projects and programs according to a given existing and forecasted capacity, then if you lack resources, that's weak planning and lack of anticipation—not a risk but a defect.

But of course, it's also a matter of perspective, of point of view. If for one of your suppliers delivering late is a defect and not a risk, it becomes a risk to identify and anticipate from your perspective as a client.

We then face different kinds of events which will, will probably, will maybe occur . . . or not . . .

We have to distinguish these events because the way we will treat them will be very different. These differences are based on their predictability (see Figure 5.3). First, we have the so-called "known-knowns." These are easy, they are our parameters, parts of the scope of our portfolio and its components,

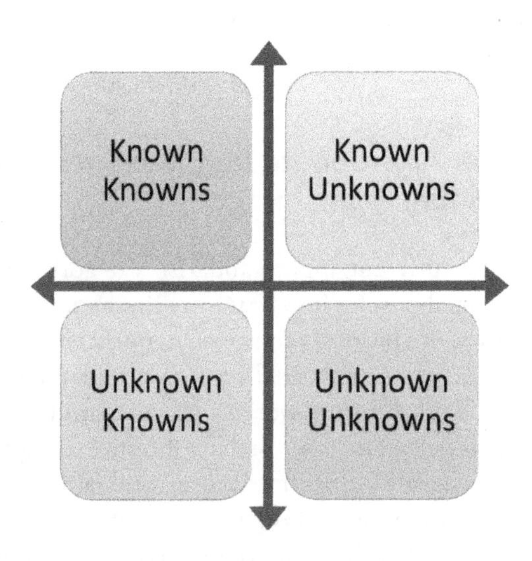

Figure 5.3 Different Kinds of Uncertainty

detailed in the project's WBS, for example. Then we find the "known-unknowns"; we know them, we have been able to identify them, but we don't know if they will happen or not . . . these are our risks. A third category of events gathers the "unknown-unknowns." We don't know what they are, and we have no means to anticipate them. That's what happens when we have a high level of ambiguity on a component. I call these events "incidentals," unanticipated incidents having an impact on a portfolio component. The last category regroups the "unknown-knowns," events we cannot define yet, but most probably will occur, and we can usually estimate that probability based on historical data. It's a probability without events clearly identified, that corresponds in fact to changes that will inevitably happen on a project or program during its execution, coming from any source or stakeholder, deliberate or emergent.

To be able to handle these various sorts of events composing our portfolio and its components, we need to construct and adapt the structure of the budgets and the associated accountability for our portfolio, but also for each component, project, or program.

Usually, a classic budget structure includes a single envelope containing everything, and eventually an additional envelop to deal with risks and other events. No need to be a certified accountant to understand that this simplistic structure (or I should say, absence of structure) doesn't allow enough control and visibility. On the other hand, we will aim at defining the simplest structure possible in order to avoid adding to the general entropy but giving us the optimal amount of information. It's then most important to develop a system that is easily repeatable and scalable throughout the different governance layers of projects, programs, and portfolios.

Here is how we should approach it:

- First we have an envelope consisting of the budget allocated to cover the scope of our initiative, covering everything as described in its scope statement documents and eventual equivalent of a WBS. Nothing less, nothing more.

  In fact this represents our ideal scenario, covering the objectives that have been formalized in the opportunity chain. Usually, that envelope is designated as the budget at completion (BAC).
- Of course, we know the ideal scenario will not happen, and some unanticipated events will occur, disturbing our very nice plans. We need then to define a budget aimed at covering the occurrence of these events and our reactions to their occurrence. That envelop will cover in fact what we called in the section above "incidentals," the "unknown-unknowns." Let's call that piece of budget the "reserve for incidentals" (RI). Often this reserve is defined by taking a percentage of the budget at completion,

usually starting with 10 percent. We'll be able to adjust that percentage later on for the portfolio and for each component.

- In addition to these, we hope to be able to identify and anticipate a certain number of events—in other words, risks, the mentioned above "known-unknowns." When constructing the risk response budget (RRB), we will have to consolidate the individual budgets allocated to each risk response plan defined for each risk identified by such a plan. That specific budget is defined by taking the expected monetary value (EMV) of the estimated cost of these response plans. Allow me to restate here that these response plans are the reactive actions which will be undertaken if, and only if and when, the corresponding risk occurs. The other actions related to a particular risk, aimed at anticipating that risk, such as avoidance or mitigation plans, have to be fully budgeted and included in the budget at completion.
- Moreover, an organization or a component team will be willing to anticipate that some changes will occur in the scope of their initiatives, and hence will assemble a management reserve (MR) dedicated to covering these changes without causing administrative complexity, creating a dedicated budget envelope to cover these eventual but certain "unknown-knowns."

With this structure in place, we have what constitutes the component budget, be it a project, a program, a project within a program, a project within a portfolio, or even the portfolio itself. Please note that this structure is intentionally slightly different from the one proposed by PMI in their standards.

In fact, the sum of these four envelops is what delimits the perimeter of accountability of the component management team (see Figure 5.4). Any decision whose impact is bounded within these limits has to be dealt with by the component manager. Any decision whose impact goes beyond these limits has to be escalated to the upper governance layer. The budget at completion is used on a daily basis to cover the costs incurred by the activities as they were defined in the management plan. When a risk which has been identified and for which a response plan has been defined, then the corresponding budget (full cost of the response plan, not only the EMV) is transferred into the BAC, baselines are updated, and the initiative moves forward. If an incident occurs, the component or portfolio manager defines the reaction plan, makes the related estimates, and covers the cost of these actions from the reserve for incidentals, transferring that into the BAC and moving forward. These three parts of the budget are placed under the accountability of the component manager, who has to decide on their usage according to the perimeter of accountability determined and bounded by the overall component budget. A possibility is also offered here to portfolio and component managers to anticipate the usual amount of change which might (and will) occur during the execution of the components. Based on an

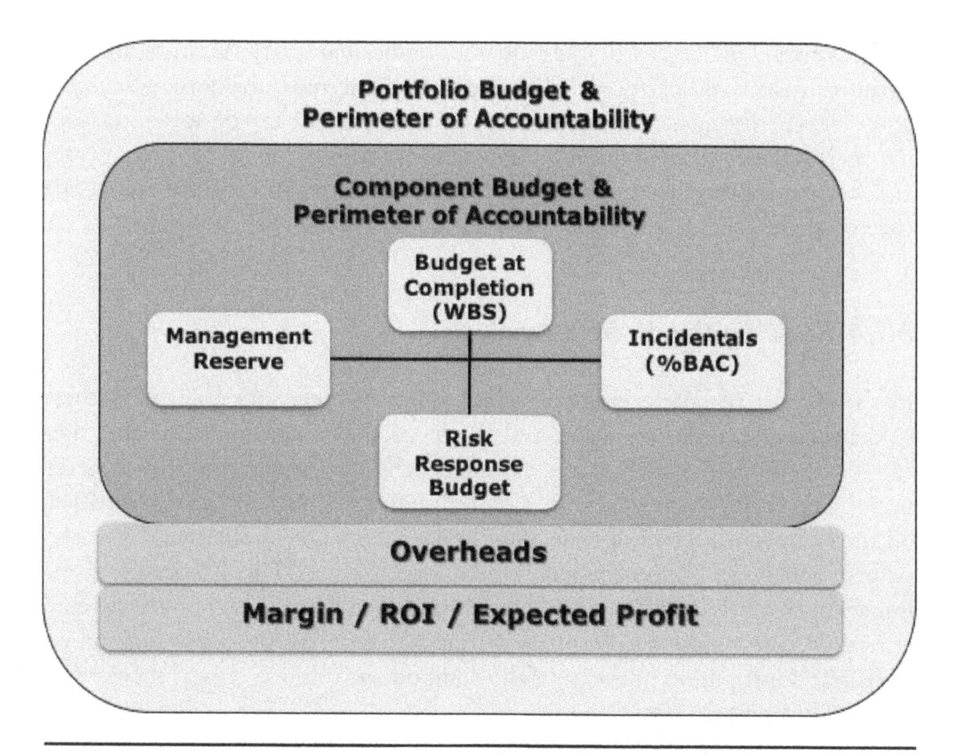

Figure 5.4 Budget Structure and Perimeters of Accountability

estimation process—often fed by historical data—one could estimate that the users or clients of a project will introduce a certain amount of change into the project, with the corresponding amount budgeted in the management reserve to be used when a change request comes, saving time and procedures in the implementation of that change and maximizing the potential added value. That management reserve should be placed under the accountability of the component sponsor (program or portfolio manager) (Lazar, 2015b).

The benefit of using such a structure mainly resides in the ability it gives to align all governance levels within the same model. That budget structure is used and constructed the same way at each level, project, program, or portfolio. This structure also provides the means to monitor the different events occurring within the portfolio and its components. How the risk response budget is consumed tells a different story than the consumption of the reserve for incidentals, and as they address events of a different nature, it's important to understand what is really happening in a given situation.

As most of the component managers will have a limited accountability bounded by their component budget (BAC + MR + RRB + RI), the accountability of the portfolio manager will integrate some additional layers. As this is aimed at optimizing the usage of organizational resources and maximizing the generation of business performance, often the portfolio manager will also be accountable for that performance (contribution margin, return on investment, or direct profit) and for making sure that the overheads (indirect costs and infrastructure) are covered as well.

5.1.4 Specificities in Risk Analysis

In the context of portfolio management, the treatment and analysis of these risks (positive or negative) will mainly be focused on their impact on the strategy, on the ability to deliver the expected benefits and generate the targeted performance, not so much (or in a more limited manner) at the project level, on their impact in terms of time, cost, and quality. The qualitative risk analysis here will prevail upon the quantitative risk analysis. The assessment of the impact of those risks will be assessed with regard to their impact on the defined critical success factors and business drivers addressed by the portfolio, and on the direct contribution of the portfolio components to these critical success factors and business drivers.

Relaying on the opportunity chain, we'll make sure to secure the delivery of the critical success factors, and the main weighting factor used to establish a risk scoring will be taken from the relative priority given to the business drivers and portfolio critical success factors in the opportunity chain, defined while using a paired comparison matrix. The aim here is to define the portfolio-level risk management strategy, establishing elements such as the risk appetite, risk capacity, and risk tolerance factors. These factors will help us in balancing the exposure to risk of our investment as regards the expected performance.

Risk capacity will be defined from the financial ability of the organization to absorb a certain level of impact of threats to the investments and initiatives, as well as the organization's ability to release the necessary resources to exploit potential business or strategic opportunities. At the portfolio level, it will be represented by the sum of the reserve for incidentals and the risk response budget.

Risk appetite, on the other hand, represents the willingness of the organization (or in fact its executives and decision makers) to take a certain level of risk and expose the organization to a certain level of liability. Usually, the risk appetite is expected to be lower than the risk capacity.

The difference between risk capacity and risk appetite defines the organization's risk tolerance. Risk tolerance is an input to establish the risk thresholds used at the components' levels to establish their specific risk management strategies—more specifically, what to identify as a high-level or a low-level impact;

what to consider as high, medium, or low probability. The closer the impact will be to the risk appetite, the more critical it will be. If the estimation of the impact of a threat enters into the area of risk tolerance, then obviously, that impact will be ranked as very high and drive the establishment of a mitigation or even an avoidance strategy. If that impact exceeds the level of risk capacity, then the initiative within the portfolio that will trigger that particular threat will have to be reconsidered, if not cancelled or delayed.

These factors of capacity, appetite, and tolerance have to be balanced, of course, with the expected business performance to be generated by the portfolio and its components. A projection of the expected performance, expressed in terms of financial profits, contribution margin, net present value, or return on investment (ROI) can be obtained, for example, through a Monte Carlo analysis, which will take the minimal estimated outcome, the maximal estimated outcome, and the most likely outcome and generate a distribution curve of the possible outcomes. This curve, put into perspective with the risk factors, will allow us to establish ideal targets of performance and aim at the components and levels of investment fitting within these objectives (see Figure 5.5).

Assessing achievability of each component, what's our level of confidence?

The balance between performance and exposure is an important factor, but not the only one to take into account when speaking of exposure to risk. There's

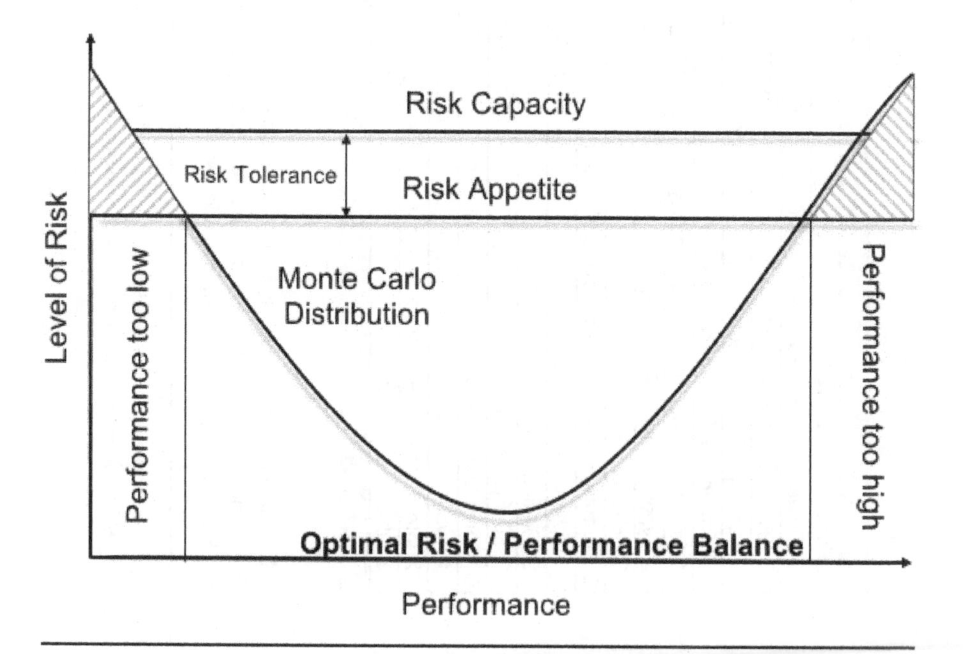

Figure 5.5 Balance between the Level of Risk and Performance

Achievability Assessment Factors	High/10	Medium/5	Low/2	Score
A. Financial Factors				
1 Project cost /total Portfolio budget*	≤ 5%	5-20%	≥ 20%	5
2 Expected Return/Benefits	≤ 3 months	3-12 months	≥ 12 months	2
3 Funding (Financial Authority)	Internal to BU	Other business unit	Outside Org.	5
B. Parameters & Constraints				
1 Resource Availability (FTE Capacity)	≥ 2:1 (200%)	2:1-1:1 (200-100%)	≤ 1:1 (100%)	5
2 Type of project (Authority)	BU initiated	Org. initiated	External/Regulatory	5
3 Schedule	Acceptable/estimated	Tight/negotiated	Inadequate/imposed	10
C. Human Resources and People Factors				
1 Spread of Resources	All internal	Internal + External	PM + External	5
2 Dedicated workforce (Fulltime)	All	most	Few	2
3 Staff Expertise/ experience	↗ Requirements	Sufficient	↘ Requirements	5
D. Complexity Factors				
1 Type of Work/Innovativeness	Known	New	Breakthrough	10
2 Interdependency of projects	Negligible	Significant	Essential	2
3 Objectives & Scope	Well Defined	Unclear	Undefined	10
Score Total				66
			Achievability	**55%**

A1 Budget includes actual and committed (pipeline)

B1 Available resources/required resources for program/project in regards of the workload (actual and committed)

C2 Multi tasking generally leads to a lower achievability

Figure 5.6 Achievability Assessment (Reproduced from Thiry, M. *Program Management* [2nd ed.]. Abington, Oxon, UK: Routledge/Taylor & Francis Group, with permission. © 2016 Taylor & Francis Group) also a factor of confidence, the perception we have regarding our ability to execute the different components themselves. The so-estimated "achievability" of each component represents our level of confidence in our ability to successfully deliver the expected outcome. For a particular component, given the current status of that component, we will look at the availability of resources and their expertise, the degree of complexity and innovation implied, and of course the related financial factors such as return on investment and payback period.

This achievability assessment has been described and applied by Michel Thiry in his book *Program Management,* but within the context of a program (Thiry, 2016). We will apply the same idea and concept, but this time at the portfolio level.

Michel Thiry's model considers four categories of factors: financial factors, parameters and constraints, human resources, and complexity. For each factor, there's a set of three subsidiary indicators. For each of these indicators, we will give a grade or score as to our evaluation of the situation of the particular component to assess. If our level of confidence in the component toward the specific indicator is high, we'll give it a 10. If our level of confidence is medium, we'll give it a 5. If our level of confidence is low, we'll give it a 2. After having graded the component toward each indicator, we can sum these grades to obtain a "confidence score," which can be easily translated as a percentage of the maximum level of confidence. That percentage is our level of confidence, our estimated probability of success (see Figure 5.6).

In itself, such an estimated probability of success is already an indicator. What will you consider as being an acceptable probability of success? It depends on your industry, on your organizational culture, and of course on your risk tolerance. Whereas a probability of success of 10 percent will be considered as a no-go for an IT development project or a business development initiative, 10 percent might lead to open bottles of champagne for a drug development program in the pharmaceutical industry, in which an average of only 1 percent of these programs reach the shelf of pharmacies and hospitals in the form of a finalized product. This probability of success is then totally relative.

In addition to the individual indicator, we can cross that achievability score with the score we obtained earlier in terms of strategic alignment (see Figure 4.10). That crossed analysis will allow us to select and prioritize the components with the highest contribution to our strategic objectives and the best probability of success (see Figure 5.7).

We will also use this achievability factor to make an assessment of the overall exposure to risk of each portfolio component and the exposure of the whole portfolio, also taking into account the evolution of that parameter over time when monitoring and controlling our portfolio.

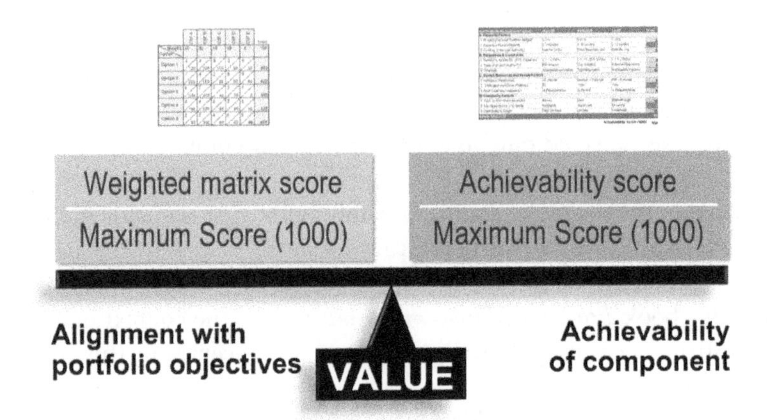

Figure 5.7 Strategic Contribution vs. Achievability (Reproduced from Thiry, M. *Program Management* [2nd ed.]. Abington, Oxon, UK: Routledge/Taylor & Francis Group, with permission. © 2016 Taylor & Francis Group)

5.2 Consolidating the Portfolio Risk Profile

The assessment of the overall level of exposure to risk within the portfolio is also obtained through the consolidation and escalation of the individual risks profiles of each component of that portfolio (Lazar, 2015c).

Risk management at the level of the individual component is indeed an essential part of the portfolio management exercise, and it is such an important aspect of the management of each component itself, that we can say that if you don't manage your risks, you basically don't manage anything. As I have already mentioned, risk is almost everything and everything is about risk, especially in portfolio management.

It's not my intent to add here the details of project or even program risk management, even if managing risks at the program level is in fact very similar to managing risks at the portfolio level. But the identification and analysis of risks at the lowest level of the portfolio is the inherent pre-requisite of portfolio risk management.

Basically, the identification of risks at the component level follows what I have described in the previous pages. The major differences will reside in the nature of the risks and the sources of their identification. We will still be using the usual categories of contextual and operational risks (see Figure 5.1) and of course apply the budget structure as described above (see Figure 5.3). In the case of a project component, the operational risk will indeed be prevalent, but the contextual risks will still keep a certain level of importance, often because they will be escalated to the upper level (program or portfolio). The budget structure proposed earlier allows us to determine which risks will have to be escalated if their impact exceeds the perimeter of accountability of the component management team—that perimeter being delimited by the limits of the component budget.

5.2.1 Assessing the Risks, Component's Risk Scoring, and Individual Profile

Often, impact and probability are the parameters considered on a project to assess the criticality of a risk. There's a long list of evaluation criteria one can use in addition to these two (knowing which ones you will be willing to use is a matter of propinquity), but the one I consider significant is proximity. Proximity represents the distance in time between when you assess or review the analysis of a particular risk and the potential occurrence of that risk. Using the proximity factor allows us to prioritize our efforts and focus on risk management of the component management team. Taking into account a variety of other factors, such as the size of the element being potentially impacted, the nature of the risk (threat or opportunity), the impacted key project parameter (critical path, CSF), and other elements, we can quanitify these aspects to elaborate an "individual risk scoring." Taken in isolation, this score is totally meaningless, but in the context of the global component risk analysis, it allows us to prioritize our focus. Figure 5.8 shows a project risk profile, using the level of probability and the proximity as vertical and horizontal axes and representing the impact of the risks by the size of each bubble (each bubble obviously represents an individual risk).

That risk profile graphical representation is easy to use when consolidating a project report for portfolio review purposes.

5.2.2 The Portfolio Risk Profile

Monitoring and controlling the risk profile attached to each project are the accountability and responsibility of the component managers, but they are essential inputs to the portfolio level.

Using these component data, we'll compute a component operational risk score, which is, in the tools I use, simply the average of all the individual risk scoring inside a particular component. This score has no meaning in itself, like the scoring of each individual component risk, by the way. They gain meaning

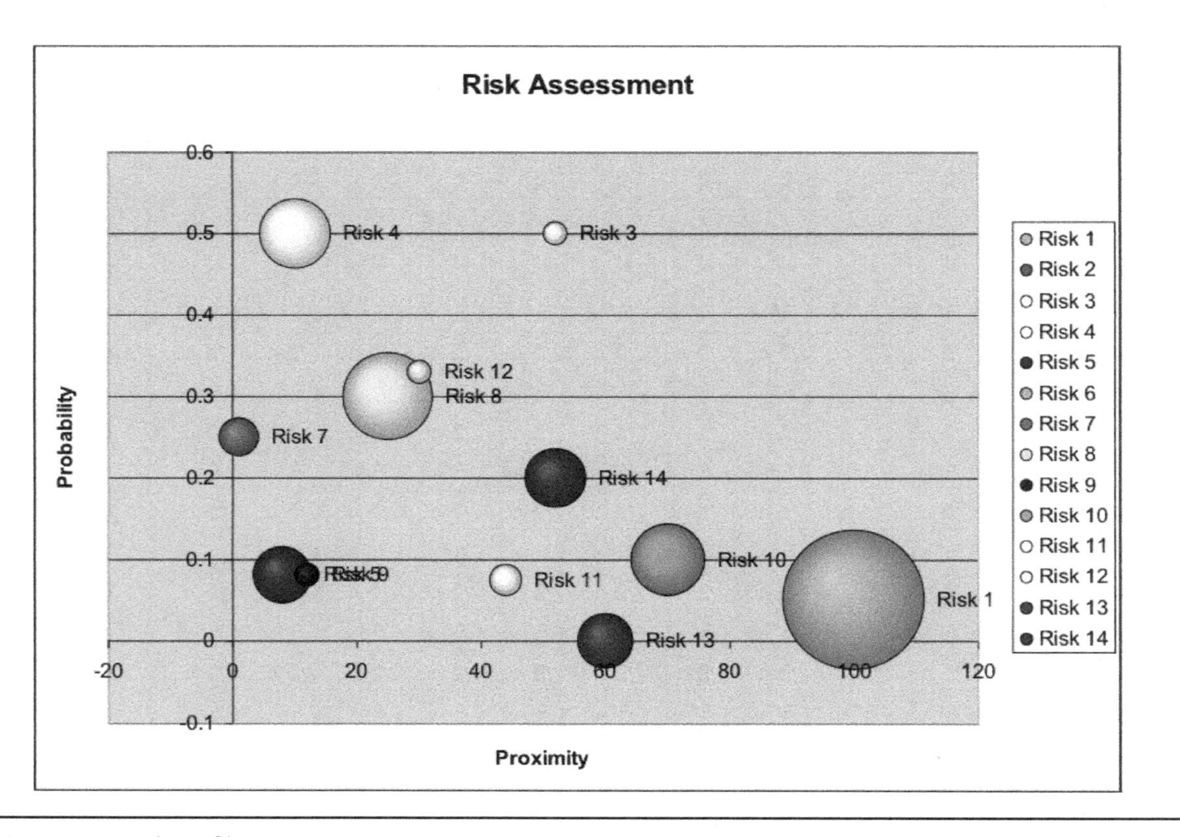

Figure 5.8 Project Risk Profile

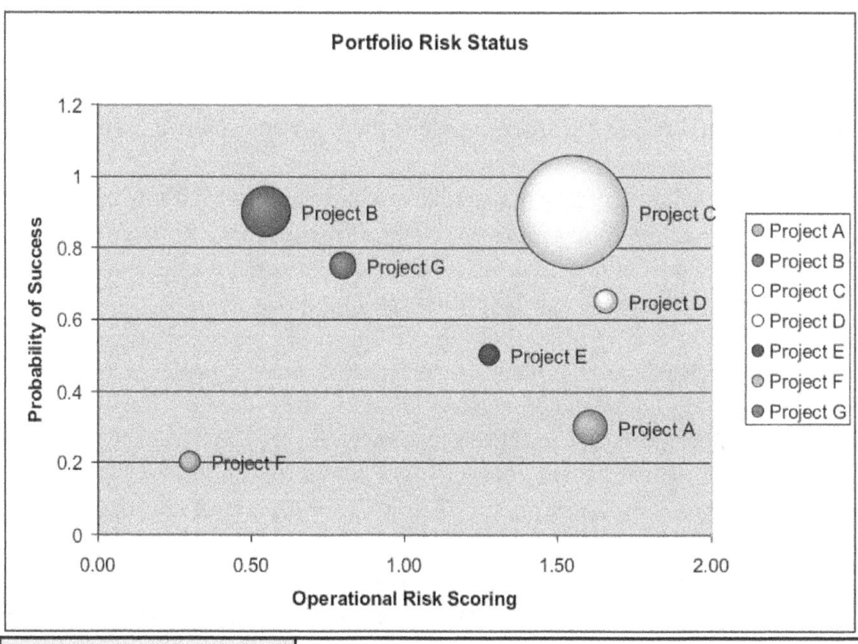

Probability of Success		
High	Secured Investment	Investment to be Challenged
Low	Investment to be Challenged	Unsecured Investment
Operational Risk Scoring	Low	High

Figure 5.9 Portfolio Risk Profile (Reproduced from Lazar, O. The Bricks for Building Your Portfolio: Risk, Benefits and Value. Portfolio Experience Conference. Warsaw, Poland. © 2015 Olivier Lazar)

when put into perspective with others: other risks within a particular component, other component risk scores within a portfolio.

We will cross this risk score with the achievability factor we obtained earlier (in Figure 5.6), this time using the size of each bubble on the chart to represent the size of the component (estimate at completion [EAC] or estimate to complete [ETC]). The portfolio risk profile then looks like Figure 5.9. In this specific example, the achievability has been called "probability of success," but it's the same factor expressed in percentages.

5.2.3 The Initial Snapshot and Component Selection

This portfolio risk profile gives us a first snapshot of the current exposure to risk of the actual and/or future components. It helps in visually determining the right balance between risky investments and routine initiatives.

We can see that components, actual or candidates, with a low exposure to risk and a high achievability represent business as usual and probably a low level of complexity, constituting a secure investment to be made. These components will usually be considered as our ground-making quick wins.

Components with a low level of achievability and a high exposure to risk are, instead, risky investments, probably representing technical or human challenges and showing a high level of complexity. These are often the major value triggers, competitive differentiators, and bearers of innovation.

The components evaluated as having a low achievability with a low risk exposure are often the elements presenting a very high level of ambiguity and would benefit from being clarified in terms of scope and objectives. Ambiguity, being a lack of qualitative information, makes it difficult to establish a clear scope statement and clearly identify the inherent risks, exposing the component to the "unknown-unknowns." A bit of additional definition effort would help in reducing the ambiguity here, and maybe the application of a program management and agile approach in this case could be beneficial.

On the other hand, the components with a low level of achievability and a high exposure to risk, while apparently being clearly defined (as shown by their high-risk scoring demonstrating a low level of ambiguity), seem to be technically challenging for the organization. Questions should be considered about the opportunity to pursue them internally or to externalize them to a partnering organization having the technical know-how—or even simply questioning their existence.

The ideal portfolio balance will not be obtained by having only components in the upper left corner. We need to take a certain level of risk if we want to generate an interesting degree of value, but if we pursue very risky elements, or if our business and organizational environment is challenging, these "risky" initiatives will have to be compensated for or counter-balanced by some easy quick wins.

5.2.4 The Evolution of the Risk Profile

The portfolio risk profile is an interesting and easy-to-read graphical representation of the configuration of your portfolio at a certain moment in time. Not only is this helpful, it also constitutes an efficient monitoring tool to control the evolution of your portfolio over time.

As we move forward in the deployment of the various components of our portfolio, their risk profile will evolve, and with it the overall portfolio risk profile.

Let's imagine we decide to launch an initiative corresponding to a component ranked with a high level of risk exposure and a low achievability, and we have decided to take on that challenge (which could be Project A in Figure 5.9). If the project goes well, as we move forward in its life cycle, we should eliminate some of the identified risks. Then the exposure to risk of that project should decrease, moving the corresponding bubble on the chart toward the right-hand side. Also, as we move forward within the development of that project, our level of confidence in our ability to achieve its objectives should rise, moving the bubble on the chart from the bottom to the top. A component in our portfolio, if developed properly, should demonstrate normal behavior on this chart by moving toward the upper-right corner from one portfolio review to the other. If you see one component going the opposite way, then you know immediately that you have a component in trouble, which requires immediate attention.

The entire idea behind this approach is to highlight the opportunities represented by the quick wins and the ones represented by overcoming the challenges of the complex initiatives, creating and delivering value and benefit triggers to the organization and its stakeholders.

Risk management is often perceived and conceived as the activity related to confront the negative events that could impact the project, program, or portfolio. Spending tremendous effort at developing risk-proof scenarios or not doing it at all because of the negative perception is just misused risk management. In fact, risk management is not so much about countering threats as it is about exploiting opportunities, finding ways to enhance the situational context and add value to our endeavors.

5.3 Exploiting Opportunities; Or the Deadly Trap of Threat Mitigation—A Matter of Mindset

It's in fact entirely a matter of mindset, and a bit of evolution theory too . . .

When hunting for prey in the savanna and trying to stay alive at the same time, our ancestors developed a natural tendency, which has been preserved by the evolution, to focus their attention on what could kill them—on threats. That's why, even today, when our environment is supposed to be safer, we still think "danger" when thinking of risk. Risk in the common language is a negative statement, and it's quite a challenge to make people admit that in project management, risk is a neutral statement.

The issue with this "glass half empty" mindset is that it leads people to concentrate their energy on constantly putting out fires, addressing solely negative events and inducing a pessimistic perspective within the organization.

We all have experienced this kind of situation, in which a major threat has been identified on a particular project, and your manager looks at you and says: "I'm totally confident in your abilities to handle this situation." In other words, your manager has just told you to face the threat and counter it.

I often illustrate this by taking an imaginary situation: You're standing in the middle of a road, with a bull elephant rushing toward you. There is a 75 percent probability you'll get killed by the elephant. Here, your manager on your project is basically telling you to stand in the middle of that road, waiting for the elephant and trying to stop it with your bare hands. That's what countering the threat means. Now, how do you feel? What's your level of motivation and engagement? And how likely will you be able to effectively stop the elephant? Probably not so much . . . But that's how most people react when identifying a threat on their projects and programs. In fact, this approach prepares the conditions and excuses for failure.

Now, let's look at the situation from a different perspective . . . If there's a 75 percent probability to be killed by the elephant, it means there's still a 25 percent probability not to be killed by the elephant—Congratulations! You've just identified an opportunity. And now, how would you feel if your manager were to tell you to work on exploiting that opportunity instead of countering the threat? Certainly, much better. And how you feel is indeed very important. It conditions your motivation, which conditions your engagement, conditioning your performance. And it's not only your performance that will be improved, but your chances of being successful in exploiting that opportunity as well. Looking at the opportunities extends the field of possibilities, multiplies options. Countering a threat is often a one way of doing things, even if quite limited, but exploiting opportunities allows us to consider a much wider scope of potential solutions and fosters creativity and innovation.

But in fact, it's the same event, it's the same outcome. The only difference is the way we look at it, consider it, and approach it, looking at the half-full glass instead of looking at the half-empty glass. That might seem a bit philosophical, and indeed it is philosophical, if not psychological. But concretely, it also helps in defining a realistic budget for your initiatives. Let's imagine that this threat with the 75 percent of probability to occur has an impact of $10,000. If we provision that impact within a risk-related budget, given the expected monetary value to consider, we're doing nothing but ensuring the loss of $2,500. Indeed, something with a 75 percent of probability will occur, no question here. And guess what, if you lose these $2,500, you will be punished because you failed at addressing a risk. Then what's the point? And a majority of organizations I've been working with still proceed in that way, if they ever address their risks at all.

Instead, the appropriate way of dealing with that event would have been to integrate it as a project parameter, identify the reverse opportunity, and start working at exploiting it. By doing so, you will preserve your project from a threat which is far from being uncertain and give you enough space and time to eventually add value by exploiting the opportunity, put yourself and your team in a positive mindset, and foster motivation, engagement, and performance.

The basic rule here is simple: When identifying a threat with a very high probability of occurring, don't consider it as a risk but as a parameter to be integrated into your scope, identify the reversed opportunity, and concentrate your efforts on exploiting that opportunity.

One might say I exaggerate with my example with 75 percent of probability. Actually, I have seen myself such things being reported in some risk registers, but indeed, 75 percent is exaggerated. I would place the limit of reversion far below that number.

Imagine another situation: you're crossing a road in very dense traffic. You have a 20 percent probability to be hit by a car while crossing the road. Will you cross it? Probably no. You've considered 20 percent as being already quite high.

Of course, it depends on the nature of the project or program, and the nature of the risk itself, but I usually start to revert threats into opportunities when they reach that level of 20 percent, which might seem like nothing, but remember it's one in five. And one in five starts to feel quite high. And it then becomes far easier to work on exploiting the 80 percent of the reversed opportunity (Lazar, 2015b).

I keep in mind the statement cited above from Peter Drucker: "Effective strategies should be focused on maximizing opportunities, and action should not be based on minimizing risks *[here he was speaking of threats]*, which are merely limitations to action." Indeed!

Chapter 6

The Fourth Pillar: Resource Demand Planning

Portfolio management as such, if summarized to its simplest formulation, is mainly about iteratively allocating and reallocating resources to optimize the generation of a certain level of quantitative performance. We have already explored the prioritization aspects of risk and strategy. Organizational agility is mainly an outcome and, at the same time, an environmental framework for portfolio management. The last fundamental element to consider is resources. We will analyze what resources, and how many of these resources, are available within the organization, knowing how these resources are currently used and will be used in the future. What future? The one corresponding to the time-frame between today and the strategic horizon we defined in earlier pages.

6.1 Analyzing the Current Capability

The first step will consist in inventorying all available resources within the organization. That activity can create a bridge, if one does not yet exist, between the PMO and the human resources department or function. What are the skill-sets and competences available within the organization? How many people do we have? What are our assets? What is our capability to deliver our initiatives and run our operations?

Two things are important here:

- Make sure that you integrate ALL of your resources into the exercise. It should include all active resources: human and material, people and equipment. We will categorize them by types, skills, and competences and issue a cartography of your resources within the organization. That cartography should map both the organizational structure and the portfolio structure, if these two structures are different (as in the case of a matrix organization). When mapping the resources spreading throughout the organization and the portfolio, we'll see the current allocation of resources (at least per skills and categories, if not individually) on the different components. This is where we also need to quantify these resources. Usually this quantification is expressed, for human resources, in terms of full-time equivalent (FTE).
- Do not confuse FTEs and headcounts. When analyzing the number of available resources, often we will count the number of people, asking them what they do, cross-checking with their job descriptions, and coming up with a number of persons—individuals or headcounts.

But when looking at the allocation of resources on the components of the portfolio, often this allocation is accounted in FTEs. But one FTE is not equal to one headcount. The reason is quite simple: an FTE represents one resource working 100 percent for one year. No one works 100 percent. People take vacations, coffee breaks, lunches, and sick leaves. It means that people are rarely present and operational for more than 80 percent of a full-time commitment—and even 80 percent is optimistic. This means that, to fulfill an FTE, we would need to have 1.2 headcount in average.

The problem this raises is that even if we allocate FTEs on our activities, we pay for headcounts and we recruit or release headcounts. Even more than that, not only do we "handle" headcounts, but behind these headcounts there are people whose performance in accomplishing their duties depends mainly on their motivation, which is triggered by their satisfaction and which triggers their engagement. So not everyone is equal there. Don't forget the human aspect behind the figures.

Once identified, these resources have to be mapped with the content of the portfolio. If you can rely upon a structured and integrated information system, it's of course easier. If you have a PMO in place, it also helps. We will go through each line of operation and ask the responsible and accountable person about what kind of resources they use, how many, and eventually precisely who.

The difficulty in this exercise resides in the time perspective. When consolidating the resource capability and allocation, we look at a calendar frame, meaning what resources and how many have been used since the last resource demand planning exercise, often January 1 of the current year. This is easy when looking at operations. It's another story when looking at projects and programs that have a different relationship with the calendar. This is where we have to look at each portfolio component, even going into the detail of these components, down to the level of the resource allocation itself, to extract the yearly resource profile of these components.

It's important here to recall some basics about project finances and distinguish the work already done, represented by the so-called *actual cost* (AC) or *actual cost of work performed* (ACWP), from the re-estimated final cost of a component project or program, the so-called *estimate-at-completion* (EAC) and the valuation of the remaining work to be done, the *estimate-to-complete* (ETC). What we're looking for here is the actual cost—the yearly figure, not the overall actual cost of a component that might span over several years.

This aspect is the perfect representation of portfolio management being the keystone of organizational maturity. Having these figures requires having in place a solid planning process on the lower level of the components. But let's be clear: That needs to be adjusted again to your real needs in terms of detail in the information and to the ability of your organization to establish these processes. Often, having a resource allocation at the lowest level of each project activity is an overburden; getting that information at the level of the work packages will be already good enough.

This data gathering is typically the kind of activity that creates a bridge between the PMO and the HR and finance departments. Here portfolio management fully plays its role as a communication-facilitating governance layer.

At this stage, it might be interesting (if we are not at the very first iteration of that exercise) to compare the yearly actuals collected with the forecast generated during the previous iteration in terms of yearly ETC for the components of our portfolio. Then we will use the gap analysis to feed the next step of the resource demand process, which will consist precisely of generating these forecasts for the remaining work in the current year and the following years, up to the limit of the defined strategic horizon, pushed forward by an increment of one year.

6.2 Anticipating the Needs in Resources for Current and Potential Components

Still in connection with the component-responsible persons, while collecting the actuals and resource allocation, we need to collect their estimation for the remaining work—the ETC mentioned above.

Again, with the operational lines of daily business, it's supposed to be pretty easy, as these parts, not being projects, are supposed to be linear in their resource allocation profile. If this is not the case, it's a good idea to investigate why. A first explanation could be that these activities have been mistakenly qualified as operations and should be considered from a projectized perspective. A second reason could be that there's a complexity factor here in the management of that part, which you have an opportunity to remove, contributing to decreasing the level of entropy of the organization. Always aim at the simplest solution. Keep in mind that the projectized perspective will introduce an additional complexity in the governance framework, so do it only if it's really necessary. But also keep in mind that not doing it when necessary is counterproductive—a false good idea.

It indeed becomes more complex with project-based components, because of the difference in their calendar time. In the resource demand planning exercise, we need yearly figures, but a project or a program can span several years.

Also, with the gathering of the ETC data, not only do we need the figures for the parts to be covered within the current year, but we need to have the figures for what has been estimated for the following years. How far? Up to the strategic horizon you have determined.

And here another factor comes into the picture: uncertainty. And uncertainty increases with time. The further you look in time the more uncertainty will be inherent in the estimates. You have to take into account that factor. How?

First, always create a reserve for incidentals (in fact, uncertainty) at the portfolio level, the same way that a reserve is supposed to have been set up for each program and project. The amount of budget that will have to be allocated to this reserve can be determined as the sum of the different reserves for incidentals of each component and sub-component, eventually adjusted at the portfolio level. Then each individual reserve for incidentals of each component will be a subset of that one.

A second element will come from applying an estimation model similar to the business development life cycle detailed in Chapter 2 (see Figure 2.5). It means applying an expected monetary value (EMV) principle to our estimates. It starts by identifying the components which have a 90–99 percent chance to be taken to their completion. Usually these are the components covering the critical success factors defined in our opportunity chain. Most probably, the estimate that has been made in terms of resources for these components have to be secured. The appropriate sizing of the risk and incidentals budget envelopes will cover the level of inherent uncertainty.

For other less critical components, and for components that haven't started yet or have not yet been approved as part of the portfolio, we will apply a weighting factor to the resource estimates.

Let's imagine, in Figure 6.1, an organization with a five-year strategic horizon. The forecast for a particular candidate component is at 10,000 man-days, and that component might start in year *N*+ 1 (next year) if approved or if the

		Years					
		N	N+1	N+2	N+3	N+4	N+5
	Confidence Level	100%	90%	60%	50%	20%	10%
Candidate Component	Initial estimate		2,500	2,500	2,500	2,500	
	Resource demand planning		**2,250**	**1,500**	**1,250**	**500**	

Figure 6.1 Resource Forecast for the Weighted Resource Demand Planning

contract with the client is signed, and for a duration of four years, meaning it will end at the final edge of the strategic horizon. The forecast to be considered in the resource demand planning exercise could then be spread as depicted in this figure (imagining that the resource allocation is spread evenly for the duration of that component).

The weighting (confidence) factor that is used has to be determined specifically for your organization.

This approach will of course not allow us to fully resource each candidate component, but as we move forward within the strategic continuum, the resources allocated to one component that has been dropped will be re-allocated to the components that are confirmed.

On project-based components, the estimates of the ETC are then collected, eventually validated and confirmed by the various component managers and the PMO.

This resource forecast is a demand-driven process; we don't consider our capability here. The aim is to see the level of resources necessary to execute all components we wish to have in the strategic plan and predict the adjustments that will have to be made, the necessary recruitments, and the competency development plans.

In addition to that, the forecast of resources has to be translated into financial terms, considering that most of the organization has evaluated an average cost for an FTE.

This bottom-up exercise is then consolidated and escalated to the organization's decision-making levels. The statements and questions to share with the decision makers are quite simple:

- "Here is our capability today," and give them an overview of the current capabilities available within the organization.
- "Here is the amount of resources used to execute your strategy as you have defined it since the previous iteration."
- "Here is the amount of resources which will be necessary to execute all of the components identified in the current configuration of our portfolio. And here is the gap between the demand and our capability."
- "Is your strategy still valid, or do you want to revise it?"
- "Have your priorities shifted?"
- "Do you approve the new financial forecast and validate it as the budget for the next iteration?"

This is where the portfolio optimization and balancing begin. Taking into account all of the evaluation aspects we have explored in the previous pages—including, of course, the data about the overall performance of the portfolio—the portfolio managers and portfolio sponsors have to consider these different perspectives and inputs to elaborate a new or updated portfolio mix—eventually a new or updated strategy, if not a new or updated strategic vision.

It means of course we also must question the previous assumptions, re-establish the critical success factors, and update the key performance indicators.

From this analysis, a new guidance in terms of portfolio planning will have to be issued and eventually applied.

6.3 Reconciling Bottom-Up and Top-Down Perspectives

The application of the newly issued guidance following the escalation of the resource demand goes through the application of the newly established plans and priorities throughout the portfolio. It's a top-down process, which is capacity driven and not demand driven as is the resource demand itself.

Often with the support of the PMO, the project teams have to realign their plans, portfolio and program managers have to review their critical success factors and the allocation of resources to the different components of their portfolios and programs, and finally operations managers have to adjust the pace of resource consumption in their business lines.

Most of the time, this reconciliation will require several iterations of bottom-ups and top-downs before a balance is reached between the strategic directives established by the top level and the tactical application of those decisions.

By the way, this is a great time to reconsider the strategic horizon of the organization, recalculating the organizational inertia and evaluating the business environment dynamic.

As part of the revision of the strategy which might be triggered by this resource demand planning exercise, reconsidering the position of the strategic horizon is an important aspect of this entire endeavor.

6.4 Establishing a Regular Resource Demand Planning Process

As with any part of portfolio management, the resource demand planning is an iterative exercise. It's to be repeated at least once a year. And every year, we push the strategic horizon by a single time increment (often one year).

The exercise of resource demand planning is the trigger of the portfolio review and analysis—its alpha and omega, start and conclusion—before relaunching a new cycle.

Typically, the iterations within what we could define as a portfolio management life cycle start with the very first iteration of the resource demand planning. Often, it's the most difficult and painful part. It can sometimes take several months to complete. In a company in which I was running this process, we used to start in April, to have some figures to present to the executives in October, having them formulate a guidance in November and reconciling the bottom-up and the top-down by the end of December. That's why it is important to have a solid and integrated information system to rely on, which will ease the exercise. I've done that with spreadsheets, and it's something I wouldn't wish on my worst enemy.

Then the portfolio has to be reviewed on a quarterly basis. Among these quarterly reviews, the first and the third will be "routine," checking the performance and the alignment of the components, reviewing the risks, and making sure everything goes as planned.

The second and the fourth will consist of deeper reviews and eventual realignments, including questioning the portfolio mix, the priorities of the critical success factors, and the strategy. If you don't obtain the expected results, it may be because your strategy needs to be reviewed, or even because you have to reposition your strategic horizon. The last portfolio review generally corresponds to the realignment triggered by the resource demand planning exercise.

These reviews serve as communication points, realignment, and performance verification. They are also opportunities to question the governance processes in place, aiming at a continuous improvement and simplification to reduce the organization's level of entropy and overall inertia.

Chapter 7

Managing Your Portfolio

Portfolio management, by its very nature, is a continuous, on-going, iterative process. There is no real temporal distinction between defining, constructing, executing, and monitoring your portfolio. It's definitely not a kind of so-called "waterfall" process, in which you would go from step A to step Z, following a predefined sequence. Portfolio management is more of an "agile" process, with permanent adjustments and measurements made during each rotation of the model and feeding the realignment executed in the next cycle. Managing your portfolio consists in fact of elaborating your strategy, executing it, measuring the achievements, and realigning what needs to be realigned (see Figure 7.1).

Many specific toolboxes and methodologies exist, detailing the inner processes necessary to effectively manage your portfolio, from PMI's *The Standard for Portfolio Management*—4th Edition (2017b) to *Management of Portfolios (MoP®)* from AXELOS (2011a), among the most known. Choosing one system over the other depends mainly on personal preferences, organizational and business constraints, and the level of organizational maturity. With a lower level of maturity, when working with my clients to implement these principles, I often start with highly structured approaches, allowing them to put in place the fundamental elements rapidly. When the maturity raises sufficiently, and the different principles are sufficiently absorbed within the organization, we can afford to introduce more flexibility and start a salutary simplification of the model, introducing and applying the various adjustments described in this book. And even if my intent here is not to redundantly step into the field of these standards, methodologies, and norms, there are some specifics introduced by the elements described in this book which are worth exploring a bit.

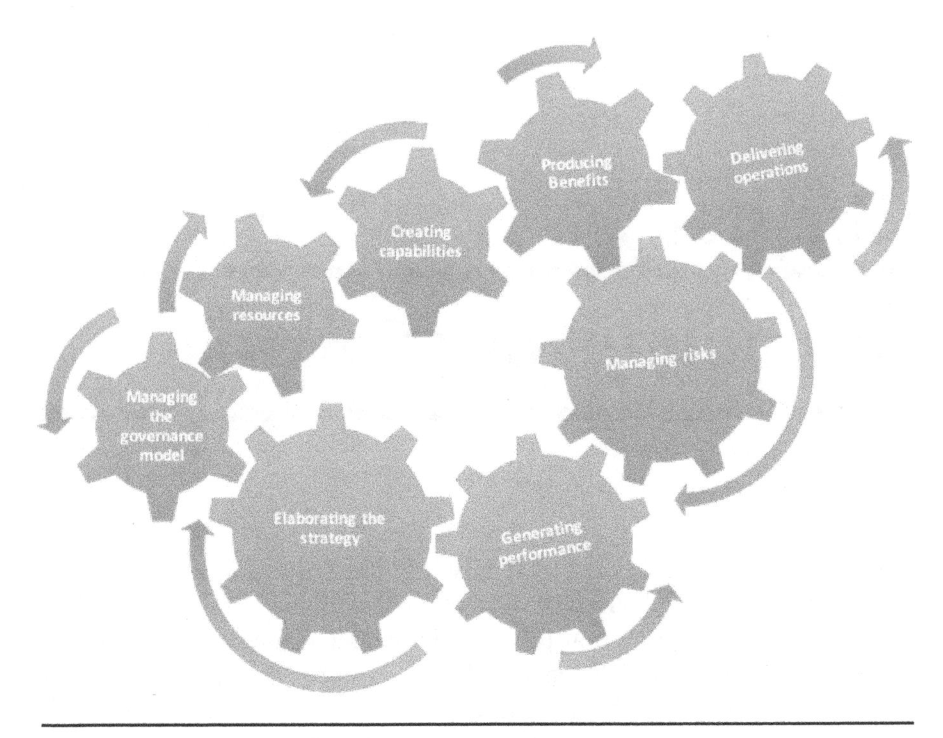

Figure 7.1 Interactions Among Portfolio Management Activities

7.1 Defining Your Portfolio's Roadmap

Defining the roadmap of your portfolio consists of elaborating the pacing of the various components (see Figure 7.2). The roadmap has to cover at least a timeframe corresponding to your strategic horizon. The components of the portfolio are expected to generate their benefits and performance within the range of the strategic horizon, including generating new capabilities (business, operational, or organizational). The expected return on investment (ROI) and payback periods of the components also have to be bounded by the strategic horizon, which will then condition the elaboration of the various corresponding business cases. Eventually, the on-going operational activities, or daily business, can exceed the strategic operation if the probability of their sustainability is low. We will then place and pace the projects, programs, and operations according, of course, to their logical sequence, if there is one, and their interdependencies, if any, but also according to the critical success factors (CSF) which have been identified while elaborating the opportunity chain (see Section 4.3.6). The idea is to spread the delivery of the critical success factors in a way that best serves the strategy and allows us to obtain the highest level of satisfaction of our stakeholders and secure their support.

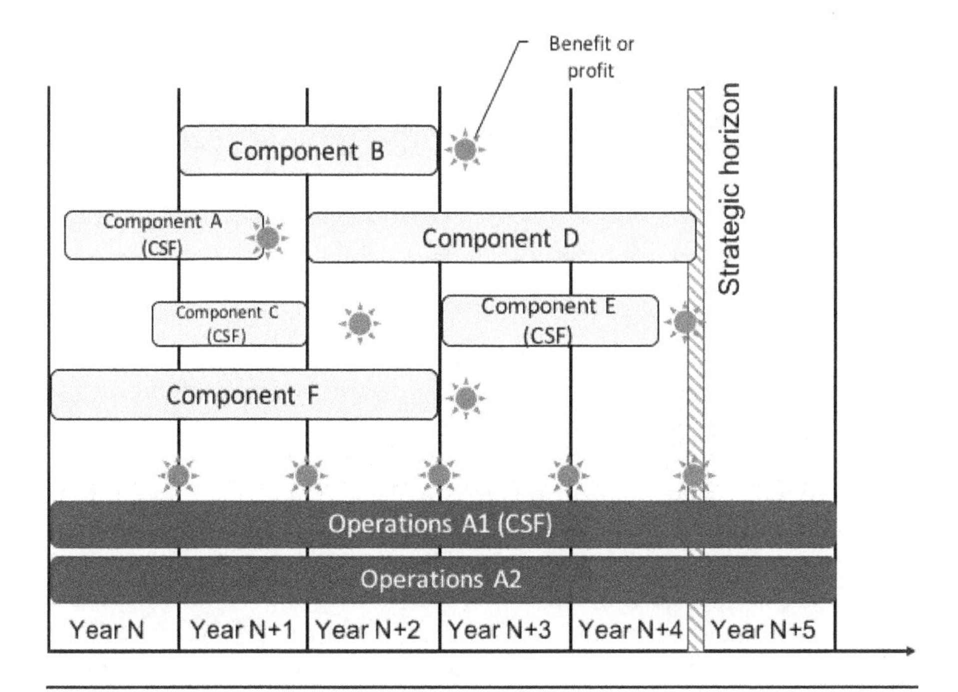

Figure 7.2 Summarized Portfolio Roadmap

If properly defined, the critical success factors should basically correspond to the Pareto principle of the value to be created within the portfolio: 20 percent of components cover 80 percent of the expected value and, eventually, 80 percent of the performance to be generated.

There are at least two approaches which can be adopted to develop the portfolio roadmap, based on the spreading of critical success factors:

- One approach will consist of securing the value and performance generation by concentrating the components covering the critical success at the early stages of the strategic timeline, planning their execution and integration as soon as possible.

This approach, which we can call *urgency portfolio planning,* will be used mainly in organizations evolving in very dynamic and complex business environments, for which the market evolution rate (see Figure 4.1) is very low. That low evolution rate shortens the strategic horizon and constrains the organization into delivering the highest value triggers as fast and as early as possible, leaving the non–CSF-related component in a "nice to have if we have time and resources" kind of category.

This approach raises a certain number of potential issues.

One of these issues is related to the pressure exerted on the organization—in other words, on the people. Often, compressing the delivery of all these critical factors in a relatively short period of time might reduce the time available to introduce periods of stability, which are necessary to allow people not only to absorb the change triggered by the portfolio components, but also to efficiently produce and measure the expected benefits. Shortening these periods of stability may lead to quickly reaching the edges of the organization's ability to absorb change, triggering potential resistance and putting the whole realization of benefits in jeopardy.

The urgency portfolio planning approach can also limit the ability of the organization to introduce adjustments in the definition of the critical success factors, also limiting the overall organizational agility. It puts the organization in a reactive rather than anticipative mode, and people might have the impression of being in a continuous change maelstrom.

If you ever have to adopt this approach, pay a special attention to the balance between risk and performance (see Figure 5.5). Generating a high level of performance very quickly can be difficult to handle from the perspective of operations, such as retail and sales or supply chain and production, causing shortages in supply, triggering a loss in customer satisfaction, and generating an

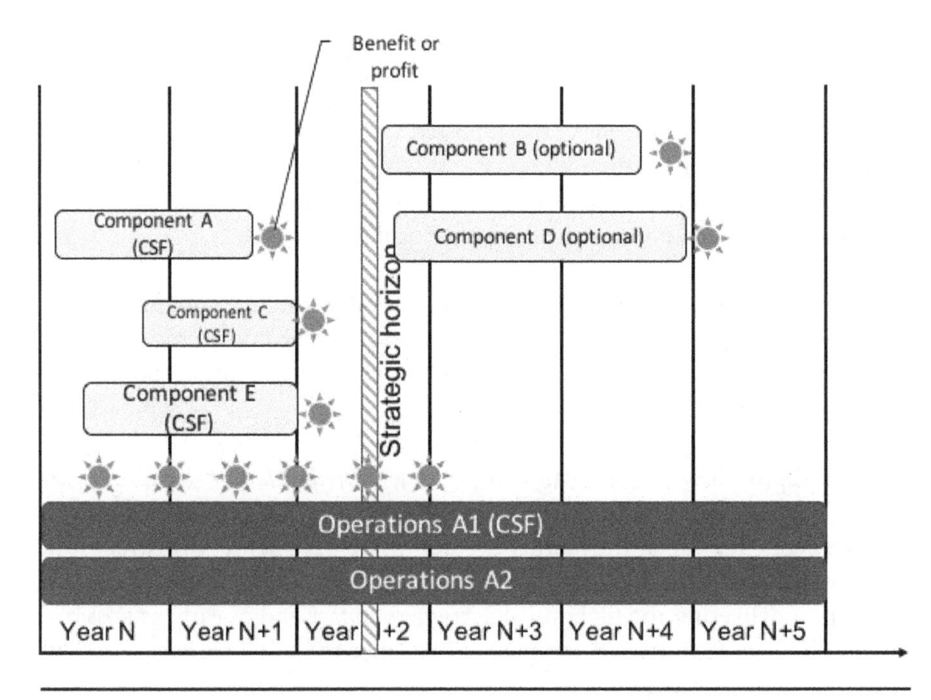

Figure 7.3 Urgency Approach to the Portfolio Roadmap

opportunity cost which might be difficult to compensate. This approach also leads to a shortening of the pace of performance measurements to maintain an acceptable level of visibility and reactiveness (Figure 7.3).

- A different approach consists of pacing the delivering of the critical success factors in order to spread the delivery of the major expected benefits along the strategic continuum, up to the strategic horizon, and maintaining a continuous flow of incremental value creation (refer to Figure 7.2). This approach allows us to facilitate the engagement of stakeholders and to ease the introduction of stability periods throughout the portfolio roadmap. These stability periods allow us both to integrate the change and benefit triggers within the organization in order to generate the expected level of performance and to measure that performance in a more effective way. Again, the risk–performance balance needs to be considered, and the long-term approach will reduce the risk of entering into the low- or high-performance zones overlapping with the organization's risk tolerance (refer to Figure 5.5). Finally, this long-term approach to the definition of the portfolio roadmap will also allow us to push the strategic horizon further, gaining in organizational agility and allowing us to reconsider the strategy on regular basis, while still respecting the change absorption ability of the organization, taking into account its inherent level of inertia.

7.2 Delivering and Managing Benefits, Value, and Performance Triggers

The roadmap will then define when each component is supposed to be executed and for how long, when the result of that component is expected to be delivered, and when the benefits and performance are to be obtained. Remember that the performance can be generated only if the organization is able to exploit the means created by the portfolio components, and this absorption is conditioned by the acceptance of these means by the stakeholders having to exploit them (the affected stakeholders). As said, a change can be considered as successful when it's not a change anymore (Lazar, 2016).

The essential role of the portfolio manager with regard to the various components of that portfolio will mainly consist of exerting sponsorship responsibilities. Often, the portfolio manager acts as the sponsor of the portfolio components, and sponsorship is a very active role. A large part of project failure is due to poor or total lack of sponsorship. The sponsor's role, though, has a very simple definition: it's the individual (or group of individuals) owning the expected benefits (Bucero & Englund, 2006). And as a portfolio manager, one has various sponsorship duties to carry out.

For project components that do not belong to a program (if they would have been part of a program, the sponsorship of these projects would have been delegated to the program manager), the sponsor owns in fact the entire project initiation process group. There's not even necessarily an appointed project manager in that step of the project management life cycle. The initiation of a project consists of defining the why and the what of the project. The sponsor brings all the necessary inputs for the project initiation, such as the business case and the portfolio's prioritized critical success factors. These inputs serve as a foundational brick in the assembly of a project charter. The portfolio manager/sponsor then owns the process of construction of the project's opportunity chain, at least down to the step "formalizing the objectives" (refer to Figure 4.15).

Owning of course doesn't mean *doing*. Often that part is covered by people with business analysis skills. It's not uncommon to see the person having done the definition work as a business analyst putting on a project manager's hat when the project charter is approved. That project charter is a contract between the sponsor and the organization, by which the sponsor is held accountable for producing the benefits and performance expected to be obtained from the result of the project. It's what connects the component to the portfolio's strategy and to the overall organizational strategy.

The charter is also a contract between the project manager and the sponsor, by which the project manager is held accountable to deliver to the sponsor the results of the project that will allow the sponsor to produce the benefits and performance expected to be obtained from that result. But it's also a contract between the sponsor and the project manager, by which the sponsor is held accountable and commits to the project manager to provide all necessary resources, support and funding, so the project manager will be able to deliver a project result that will allow the sponsor to produce the benefits and performance expected to be obtained from that result. Because the portfolio manager is the sponsor of the projects that are directly under his/her level of governance, without having the program level necessarily in place, the portfolio manager has then to track, follow, and control the overall performance of the project in terms of process assurance, making sure the project is managed according to the set standards.

The portfolio manager also has in his or her sponsorship role to make sure that the functions covered by the result of the project in the portfolio's opportunity chain are effectively covered and meet the key performance indicators' levels, within their range of flexibility for the given criteria (see Section 4.3.6), especially if the function in particular has been defined as a critical success factor for the portfolio.

The sponsorship principle to apply to a program component is quite similar in terms of accountabilities and responsibilities, and the contractual nature of the program charter (or mandate) is similar to the one established for a project. The main difference here is related to the differences in nature between a project and a program. Here the result is less tangible and less immediate to measure, as a program is expected to generate business benefits or organizational capabilities. As a sponsor, the portfolio manager mainly has to validate that these new capabilities are properly integrated within the organization, that the eventual factors for resistance have been addressed, and that these capabilities permit the organization to generate the expected level of performance obtained from the operations performed by using the said capabilities.

As part of this portfolio manager/sponsor role comes, of course, that of allocating the necessary resources and eventually reallocating these resources from one component to the other, based on their individual status and their relative priority. While allocating resources and overseeing performance, the portfolio manager is also the one authorizing the execution of these various components, sometimes validating the different phase gates and also sometimes putting a component on hold, postponing it, or simply terminating it if there's any indication that the component will not deliver what it's supposed to deliver or if there's any indication that the result will not contribute to the organization's or the portfolio's strategies.

When securing and measuring the realization of the expected value, the portfolio perspective allows us to consider a much wider scope than that of a project or a program taken in isolation. The aim of a portfolio being mainly to generate performance, the role of the portfolio manager will be to fill the governance and management gaps that might occur during the periods of absorption of the component results and the inevitable latency between the moment a component has delivered its result and the moment the benefits or capability created can effectively generate the expected performance. This latency is necessary, and it's the role of the portfolio management to manage it while keeping track of the evolution or the propagation of the components outcomes. Again, if we do not allow for these times of latency or stability, the benefits will never be produced, and the capabilities will never be able to generate the expected performance.

7.3 Assessing Performance

The performance of the portfolio as a whole and the performance of each component have to be considered from different perspectives:

- The absolute tactical and quantitative performance, measured, for example, in terms of earned value data and conformity of the results of the components to their specific key performance indicators (KPI)

- The strategic, qualitative performance, measured by looking at the coverage of the portfolio's critical success factors and the organization's business drivers

The most important is always the qualitative. Without the qualitative, the quantitative performance is useless—that information system you have delivered on time, on budget, and on scope but that nobody uses. A very efficient waste of money and resources.

In addition, the performance we seek to obtain through our portfolio, even if mainly quantitative (profits, income, ROI, productivity, etc.) has the particularity of being an aim on its own. That kind of performance is always a consequence of a well-defined and properly executed strategy. It's a symptom of a good strategy.

Then, as in medicine, we'll look at the symptoms to assess the health status of our patient—our organization—but treating the symptoms is not the aim of medicine, as fixing performance flows is not the aim of portfolio and corporate management.

Let's give a look at those quantitative symptoms first, as they are the easiest to analyze.

At the portfolio level, we will look first at the consolidated performance indicator coming from the execution of the components. The simplest expression of these indicators can be taken from earned value management (EVM), looking at variances (cost and schedule variances), performance indices (CPI and SPI), the estimate to complete (ETC)—both global to the component and yearly as described in Chapter 6—and of course the actual cost and the estimates at completion. We'll of course make sure not to use the SPI calculation based on cost, but the one based on time, using as our basis the earned schedule figures instead of the earned value figures (Lipke, 2012).

To add to these performance metrics, we'll have to monitor specifically at the portfolio level some of the financial indicators, at least the ones that serve to elaborate the business cases of our portfolio. Among these financial indicators, the most relevant ones are also quite classic, such as revenue, net present value, contribution margin, or ROI. One last sort of business performance indicator will be related to the usage and allocation of resources, tracking the overall capability allocated to the portfolio for a given year (see Chapter 6), the part of this capability allocated to each specific component—also for a given year—and the unallocated part (nothing obliges you to allocate your full capacity at once). We'll track the evolution of these indicators at each portfolio review and document this evolution in the appropriate tool (often a spreadsheet from a well-known software company works very well).

7.4 Tracking Risks Within the Portfolio

As per tracking quantitative data, we also need to look at the sort of data which will allow us to make the bridge between quantitative and qualitative—this means risk.

Relying on the budget structure described in Section 5.4.1 (see Figure 5.4) to establish the need for reserve for incidentals and a specific risk–response budget, I've found it interesting not only to track the evolution of the consumption of these reserves in absolute numbers, but also to look at their relative consumption for each component.

As a reminder, the budget at completion (BAC) represents the scope of the component, what has to be done (known-knowns). The reserve for incidentals (RI) is defined to face the unpredictable occurrences (the unknown-unknowns), and this reserve is adjusted according to our ability to properly identify risks (known-unknowns, see Figure 5.3) and construct the risk response budget (RB) from the budgets allocated to the different risk responses defined and planned in front of the potential occurrence of identified risks.

Let's imagine a scenario occurring on one particular component of our portfolio: When looking at the consumption of the BAC, we see that 30 percent of that budget has been consumed. Normally, 30 percent of consumption, if the component goes as planned, should correspond more or less to 30 percent of progress in the development of that component.

Now, we look at the reserve for incidentals and see that 80 percent of that reserve has been used. There something here that should catch your attention. An overconsumption of the reserve for incidentals, relative to the BAC, means that a lot of events, actually too many of them, which you have not been able to anticipate occurred. This means that the orientation of your risk identification was wrong, and your risk analysis has to be revised. That's not good news indeed, but the good news is that your component is still within its budget perimeter; it's not overbudget yet, and there is still time to react and do something within a reasonable horizon, instead of discovering the problem once it has happened and entering into the biases and flows affecting the decision making in panic mode.

Let's imagine a slightly different scenario now for our component. We're still at 30 percent of the BAC and on progress. But looking at the risk–response budget, we see that we have consumed 80 percent of that reserve. Again, the story told by the relative consumption of these budget envelops is interesting. The overconsumption of the risk-response budget means that a lot of events, again too many of them, which you have considered as uncertain finally occurred. Someone can have bad luck, of course, but a ratio of 80 to 30 is disproportionate.

It means, in fact, that a lot of these events which have been considered as uncertain should have been considered as parameters of our component and integrated in its scope statement. So, the scope statement has to be reviewed. Again, not good news, but good news considering we're not overbudget yet on that component, and we have anticipated and avoided the panic mode, which is the aim of proper portfolio management.

Another aspect of risk to consider comes from the consolidation of the risk profiles, as mentioned in Section 5.2.1. We'll look at the evolution of the individual risk profiles for each component, as seen in Figure 5.8. The different risks

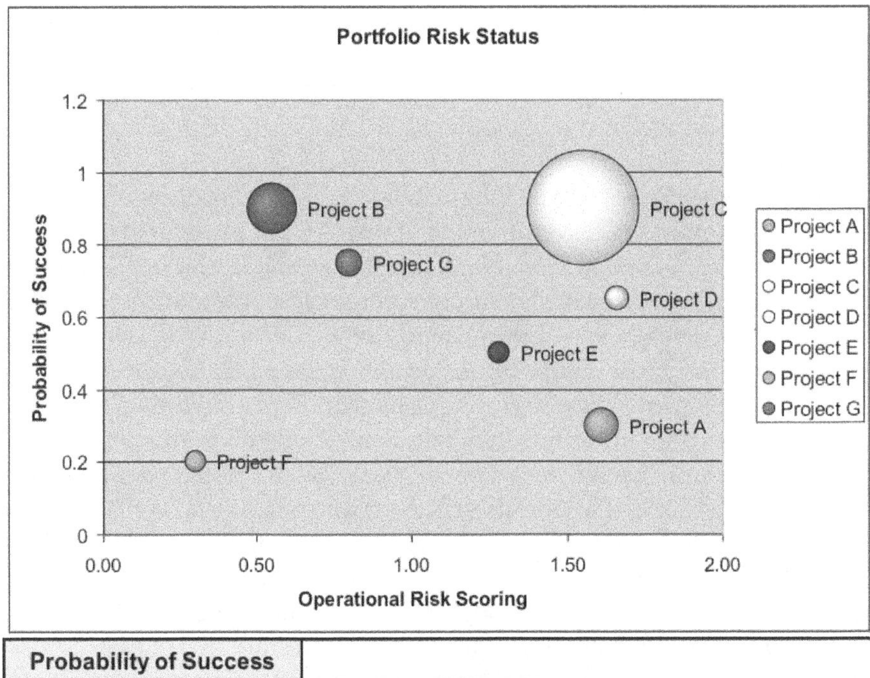

Probability of Success		
High	Secured Investment	Investment to be Challenged
Low	Investment to be Challenged	Unsecured Investment
Operational Risk Scoring	Low	High

Figure 7.4 Reminder of the Portfolio Risk Profile (Reproduced from Lazar, O. The Bricks for Building Your Portfolio: Risk, Benefits and Value. Portfolio Experience Conference. Warsaw, Poland. © 2015 Olivier Lazar) (repeated from Figure 5.9)

represented should normally decrease in size and/or in height if some mitigation plans have been put in place within the components. This is the same for the risks escalated at the portfolio level but dealt with at the portfolio level. Definitely if a risk, as it's coming closer on the proximity axis, is going up on the probability axis and gets bigger in size, we don't need much more digging into the details to figure out there's something wrong happening with that component.

Looking at the overall portfolio risk profile, detailing the risk exposure and achievability of each portfolio component (see Figures 5.6 and 5.9), we'll be able to assess the general status of each component. The portfolio risk profile, as shown in Figure 7.4 (repeated from Figure 5.9), gave an instant snapshot of the exposure to risk of the portfolio, allowing us to prioritize the components in order to achieve a certain balance in the overall exposure to risk. It's also interesting to track the evolution of that risk profile on a regular basis, which can even be monthly and not quarterly, as are the global portfolio reviews.

Let's imagine we decide to include and execute a candidate component which had been assessed as highly exposed to risk and low on achievability, meaning a risky business positioned on the lower right corner of our graph. Normally, if everything goes on time and smoothly on that component, as we move forward in its life cycle, the exposure to risk should decrease, and our level of confidence should increase, moving the project toward the upper left corner of the graph. That's the ideal and normal scenario. But if we see that this component is moving in the opposite direction, again there is no need to dig deeply into figures and reports to understand that this component presents some difficulties.

One last dimension of risk to consider sits at the portfolio level itself: The risks directly identified or escalated from the different components because their impact exceeded the perimeter of accountability of the component managers, those risks that have been evaluated with regard to their impact on the realization of the portfolio's critical success factors and the organization's business drivers. We'll need to keep track of their evolution, in terms both quantitative and qualitative of their impact on the critical success factors and business drivers, or in terms of their financial impact on the business case of the different components or on the portfolio itself. The impact of these risks on the critical success factors or on the contribution of the portfolio components to their realization or the realization of the organization's business driver might lead to reconsidering the strategic alignment of the impacted components, if not the strategic alignment of the whole portfolio.

7.5 Ensuring the Strategic Alignment

Of course, maintaining and even securing the alignment of the portfolio with the organization's strategy is more than essential.

Several simple steps can allow us to control and verify that. The first of these steps consists of looking at the critical success factors. Are they still critical? Isn't there one which was not critical yesterday, that should be considered as such today? Then we'll look at the components related to these critical success factors and to the business drivers and check their status in terms of performance and progress. Using the same evaluation model that has been used in the construction of the opportunity chain, when formalizing the objectives after having identified the needs (see Figure 4.10), we'll update this assessment and verify its validity.

And as we move forward within the strategic continuum, we'll come back to the opportunity chain and follow it backwards. Are my components covering the functions they have been designed to cover, according to the corresponding key performance indicator, reaching the level for the criteria within the defined range of flexibility? If so, then the functions should have been covered. Are these functions designated as critical success factors? Have I covered each of them, especially if the portfolio takes place within a roadmap constructed upon the principles of urgency portfolio planning (see Figure 7.3 and Section 7.1)? If so, we might have achieved most of our objectives, at least those that generate most of the targeted value. And if our objectives have been achieved, then the needs of our stakeholders have been responded to, and we can assume that their expectations have been addressed. If these different layers can be ticked, then most probably the portfolio has a good chance to have created a certain level of value, within the limits of the organization's strategy, that value being possible to demonstrate through the tracking of the opportunity chain, the verification of the KPIs, and the validation of the CSFs.

7.6 Realigning and Pushing Your Strategic Horizon Forward

This exercise of regular performance measurement is mainly aimed at reconsidering, validating, or questioning the strategy. If ever the expected results are not obtained, then there are two possible sources of any deviation:

- Either the content of the portfolio needs to be realigned toward the strategy, and then the decisions in regards of recomposing the portfolio mix have to be made. These decisions might imply cancelling some components, launching new ones, and reprioritizing the ones that will be maintained, and from this reprioritization, resources might also be reallocated. And the realigned portfolio has to take place within the strategic continuum of the organization, going through all the different steps we have explored until now.

- Or do we have to realign the strategy itself? Reconsidering the plans is also part of the portfolio management activities. If the indicators are not as good as we could have expected, then maybe the baseline has to be reviewed? It can span from a complete reformulation of the strategy to a minor readjustment. The basic questions to ask start with questioning and analyzing the business environment. Has it changed? Have I gained additional visibility, or has that visibility decreased? Was my strategy efficient and effective, has it allowed my organization to shape its business environment and develop that unique competitive advantage the way it was supposed to be? If so, what next?

Each iteration of the portfolio management exercise, especially the part about resource demand planning, pushes the strategic horizon by one increment of time, usually a year. But a successful strategy, by allowing the organization to shape, create, or even influence its business environment, should also allow us to expand the strategic horizon. If at any moment, having a strategic horizon of four years, for example, you realize that you can extend it to five years, it's a sign of a successful strategy. If it goes the opposite way, and you have to shorten your strategic horizon, that's a first sign that you're losing control, and you should reconsider either your strategy or your portfolio mix.

Anyhow, you will never catch your strategic horizon. It's like running after the sunset—it's moving away from you as you have the impression you're getting closer to it. And the aim is not to catch it. The aim is to write the story of your organization, giving a purpose for these people to gather and contribute to create value one for the others and create a momentum of sustainability for the organization. As long as you can keep up this momentum, your organization will be successful, people will be engaged and motivated and will keep running after the horizon, from one portfolio management iteration to the next.

7.7 Prepare the Next Iterations of Portfolio Management

Triggering the next cycle of portfolio management consists of keeping that momentum and maintaining a favorable dynamic. It's a leadership exercise, a motivational endeavor. As you progress throughout these iterations, it is of major importance to consolidate on the previous achievements, including process-wise, recycling the information, especially on estimations and risks. Establish clear and simple procedures, always look at reducing entropy and inertia and increasing organizational agility. Automation and technology can help. Harvest the potential benefits. The more you can automate all the processes we went through, the more efficient and easily accepted will be the integration of portfolio management within the organization. Keeping it as transparent and as simple as possible will allow the very practice of portfolio management to evolve itself and adapt to the variations of the level of maturity of the organization and the evolution of its environment. And here also, the evolution is rapid and hard to predict.

Chapter 8

Evolution and Future Developments in Portfolio Management

All of the principles and fundamentals of portfolio management and, in general, the overall integrated framework of strategy, portfolios, programs, and projects are constants that I don't foresee evolving extensively in the years to come. As discussed earlier in this book, the emergence of so-called "agile" practices has basically just put a label on commonsense and already widely applied behaviors, as in the play of Molière, *Le Bourgeois gentilhomme*, in which M. Jourdain, the main character, was using prose without knowing anything about it. But one thing will change for sure, and that change is already here—how we will operate this governance framework and how will we position our roles as project, program, or portfolio managers and as project management officers (PMOs).

The current trends in the expansion and development of digitalization and artificial intelligence raise some key questions about the future of our related professions and practices.

One thing is clear: Everything that can be automated will be automated. And it will go far beyond what we know today regarding current project management information systems. The development of artificial intelligence in terms of data treatment completely reframes the way we can analyze the information that we collect. Indeed, the amount of data is growing, and even if we manage to maintain a relative simplicity, as I have tried to encourage you to do in these pages, it's still a lot—so much that analyzing the whole amount of data generated at the portfolio level, especially if we talk about high-level portfolios, makes it impossible to digest these data within an acceptable timeframe and with an acceptable level of accuracy. The application of so-called "big data" is therefore not only desirable but mandatory, as the constant acceleration of the business environment dynamic and the increase in business complexity has not only increased the amount of data generated to be processed but has also drastically shortened the timeframe within which these data have to be processed. Therefore, there's a natural tendency that's important to resist—that is, shortening the strategic horizon—which is a deadly pitfall that should be avoided.

We can now honestly say that the domains (if not the practice) of project, program, and portfolio management are quite mature. PMI has published the 6th edition of their *PMBOK® Guide* (2017a), their program and portfolio management standards are now in their 4th edition (*The Standard for Portfolio Management* [2017b] and *The Standard for Program Management* [2017c]), and many other similar organizations (such as IPMA® and AXELOS Limited) have also released their own versions of these governance layers. The process structure of these governance layers is then pretty well defined, and is, therefore, ready to be automated.

And again, everything that can be automated will be automated. All of these project management tasks, such as scheduling, identifying, and analyzing risks; estimating costs; conducting procurements; consolidating performance reporting; and compiling various indicators, will be automated sooner or later. And that later looks like tomorrow. Yes, all of these activities, usually allocated to PMOs, will be performed by systems.

One might say that some of these activities, such as risk analysis, are too sensitive and blurry to be automated. That would be wrong. There are limits to what can be automated and what cannot be, and we used to think that things which require a certain level of intuition and sensibility, in other words requiring the ability to manage complexity as opposed to complication, can never be automated. It appears that this is wrong. Since no human being can now beat a computer at chess, it is clear that a computer can't be beaten at handling complication. Complication is mainly a sum of data to be treated through a clearly detailed process. It requires calculation and memory muscle. As in chess, if a system can remember and recall more wining combinations of the chess board for the current configuration of that board and can do so faster than a human, it can't lose. That's what happened when IBM's Deep Blue beat Garry Kasparov in May 1997.

But complexity is different—it's contextual, it's ambiguous, it's changing. It requires thinking agility. No way to introduce thinking agility into a computer . . . until Google developed AlphaGo, a system that beat a game of Go champion. Then the limit of what could be automated shifted—toward us. It showed that a lot of professions which we thought as being AI-proof were in fact quick and easy targets for automation and disappearance—lawyers, doctors, financial analysts, surgeons, travel agents, executive assistants, and . . . PMOs and project, program, and portfolio managers! Scared? Good!

Actually, I may have exaggerated a bit. The situation is not that bad. But indeed, our professions and roles will change and evolve—at least in the gathering, collection, disposal, and interpretation of data and their transformation into information, into an actionable decision-making trigger. These activities will be automated, and it's a good thing. In that sense, artificial intelligence will ease our lives, even raising the accuracy of the data provided, given the ability of AI to process a global set of data and extract rules of exceptions from the whole data population rather than having to extrapolate from the treatment of a limited sample.

This global treatment is what allows us now to really extract the exception from a yet unknown rule, to extract the anomaly from the global pattern.

I've worked with and seen systems highlighting correlations between events among more than one million factors in the aerospace industry, or extracting really unusual behaviors from a set of banking transactions—wherein each of them is almost an exception in itself—to find and fight frauds, or, using the same system, predict the evolution of the stock exchange market (by definition irrational) or detect causes of a defect in the production of steel. Statistics would not allow the achievement of these results. But big data processing using artificial intelligence makes it possible by reversing the usual model, wherein we first formulate hypotheses (H0, H1, H2, etc.) and then test these hypotheses by observing a (hopefully) statistically significant sample. With big data processing, we can observe the entire set of data as a whole and extract rules and patterns from which exceptions will naturally appear on which we'll focus our attention instead of having to manage the general rule first. We can then free our time, resources, efforts, and energy to focus on what really matters. It's the ultimate application of the Pareto model.

The inclusion of artificial intelligence into our project management world will allow us to focus on the real value of creative decision making and risk taking. What is the role of a manager (whatever adjective or noun you add to that designation, and whatever is the level and scope of accountability)? It's about making decisions within a certain accepted (if not acceptable) level of risk. And that no machine (so far) is even close to being capable of.

Some definitions of artificial intelligence state it as being the ability for machines to make decisions. Actually, machines don't make decisions. They can support decision making and feed it, but they can't make these decisions, and they won't for some time to come.

Automation in project management will indeed make PMOs obsolete in their traditional roles of data processing and central points of information spreading. And project managers or portfolio managers will no longer be relevant in processing these data and information, but they will remain relevant, and in a more demanding and strategic manner than ever, in their ability to make informed, conscious, and risk-aware decisions.

PMI has developed their model—the Talent Triangle®. In my personal opinion, this is one of the most important developments they have made since they put the PMP® credential on the market. The Talent Triangle® states that the project management competency at large is a combination of three dimensions: strategic awareness, what they call "Technical Project Management" (which is nothing more than organizational skills), and leadership competences. The technical part of this model describes the tools, techniques, and processes that will become automated, making the other two dimensions of strategic awareness and leadership competences even more prominent, and then transforming the organizational skills into decision-making and risk-taking abilities. These are the aspects to develop in the set of competences of future project, program, and portfolio managers, which the organizations will need with the exponential acceleration of the business dynamic and increase in business complexity. These increases and accelerations being multiplied, and not only amplified, by the extended abilities brought about by artificial intelligence and automation, are here not to replace us but to complement us, giving us the opportunity to create more value and develop that value in even more sustainable ways.

Chapter 9

Conclusion

9.1 Portfolio Management as an Organizational Maturity and Agility Trigger

After everything we've discussed in these pages, one final question remains: What is all of this about? After all, most organizations are doing perfectly (or almost perfectly) well without having all of these principles in place. Some even purposely choose not to put these principles in place, their executives saying they want to preserve the pioneer's start-up mindset. Isn't that at the end something any business owner, accountable to a profit center, is doing naturally? Indeed, at least to a certain extent.

The fact is that putting an organization—a gathering of people—in place and making use of a certain number of resources does not entail the same kind of responsibility that it might have decades ago. With a growing scarcity of resources and greater need for showing responsibility toward people on the part of organizations, the need to use these limited resources in a way that creates real value is more important than ever, and how you treat people and consider them in the value creation process is what constitutes the real competitive factor in a defined market today.

The kind of product you make, or even how you make it, does not provide a competitive advantage anymore. The average lifespan of an innovative breakthrough is, at best, three months. This means that whatever innovative product or service you put on the market, and no matter how innovative your way of producing or delivering it is, it will, at most, take three months for the first competitor to put something similar—often better and cheaper—on the market. Market explorers are rarely the ones who are the most successful.

The competitive advantage will derive from the ability to attract and retain the right people (employees AND customers)—by creating this particular bond, this relationship of trust and engagement. It might seem a bit idealistic to say that, but actually, there is a business case behind trust and ethical leadership. According to the Trust Barometer from Edelman (2016), 68 percent of people (clients) buy products and services because they come from a company they trust, and 37 percent are keen to pay more for the same product or service if it comes from a company they trust. According to Gallup in their study, "State of the Global Workplace" (2013), productivity increases by 21 percent, turnover decreases by 65 percent (yes, 65!; when knowing the cost of a new hire, it's enormous), and the number of quality defects decreases by an average of 41 percent when employees are engaged.

The only way to create that kind of engagement is to demonstrate the ability of the organization to create real and tangible value for its members and stakeholders, and, in addition to that, ensure that this value is sustainable over time (remember Figure 4.6).

A structured way of using the available resources (and only the necessary ones) is required, along with the ability to transform these resources into concrete elements. A long-term strategy that will guide the organization along the path to developing sustainable value, while still keeping enough flexibility to adapt to a constantly changing environment whose transformation rate increases exponentially, needs to be defined.

The sort of portfolio management approach we have touched on here is about providing the following:

- The organizational maturity that is necessary for integrating the different levels and components of the organization into a single framework, embracing the whole of it rather than considering its individual components and providing the means for communication and collaboration—both horizontal and vertical. This can only be achieved by a rationalization and simplification of processes and organizational charts and by being sure to measure what we do more than how we do it. Compliance and conformity, as important as they can be, can very quickly create redundancies within the organization that trap them in the deadly pitfalls of entropy and inertia.
- The required level of agility, which enables anticipation rather than reaction and respects the ability of the people within the organization to absorb a certain level of change and ensure the smoothest and most efficient integration of the means aimed at executing the strategy.
- Finally, the ability to formulate a strategy that envisions the purpose of the organization, that serves as an engagement and adherence factor for the

people involved, and that can also translate this conceptual vision into a concrete plan with tangible outcomes that people can rely on.

Organizations that will survive are not the biggest or the richest. They are the ones who will be able to foresee the changes coming in their environment and eventually implement these change themselves before someone else does it (thereby making them obsolete), and who will develop the means to adapt to those changes, respecting their clients and employees, and not only play an obvious economic role but also assume a societal responsibility in how they optimize the use of their, or our, resources.

This is what all of this is about.

9.2 Developing the Appropriate Mindsets

It's indeed a matter of mindset. It's a way of perceiving and transmitting things.

First, the productive type of mindset is the one to adopt rather than a defensive one, which is often in place within struggling organizations. The productive mindset induces transparency, communication, and collaboration. It focuses on the common development rather than on self-centric and individualistic growth, which often leads to the multiplication of interfaces and small closed boxes of accountability (remember Figure 3.2).

Second, the agile mindset, which perceives change as an opportunity instead of a threat, allows the harvesting of benefits and exploiting of the energy transmitted by change. Changes happen. With or without us. Therefore, it's always better to adapt than resist. And if the environment becomes too unfriendly, then maybe it is time to reconsider our direction and try to find new ways, along with new destinations and new ways to reach them.

Also, speaking of opportunity, and perhaps most importantly, it's key to adopt a positive mindset. Value doesn't reside within your ability to counter threats. Countering threats limits your horizon, decreases your number of options, and consumes a tremendous amount of resources, often with limited results. Look at the positive side of things, face and accept the facts, and develop the actions that are necessary for exploiting the opportunities you have thus created.

9.3 Benefits of Portfolio Management

Developing a mature portfolio management integrated framework will not just allow you to manage resources and make plans. It's an organization-wide improvement factor. If you ever want to put these principles in place within

Achieved Benefit	Principles & Tools	KPI & ROI Measurement
Enhance Visibility on the Portfolio of Projects	• Portfolio Management • Program Management • Opportunity Chain©	• Resource Allocation Plan • Contribution to Strategic Objectives • Productivity Increasing • Profitability Increasing • Consolidated Performance Indicators
Customer & Stakeholders Satisfaction	• Value Management Expectations ➔ Needs ➔ Objectives	• Less claims and complains • Less change requests • Increase in Productivity • Higher "Success Rate" • Better overall Quality (products and processes)

Respect of Time and Cost commitments	• Project Management
	  • Time Management
	  • Scope Management
	  • Cost Management
	  • Risk Management
		• Measurement and Analysis of Cost/Time/Scope/ Quality Variances
		• Earned Value Management
		• Fit between Plans and Execution

Better Projects' Profitability	• Planning and Estimation techniques • Earned Value Management	
Increase Organization's level of Maturity (PM and overall)	• Project Manager Competency Framework • Portfolio Management • Program Management • Project Management • PMI Certifications	• Level of Organizational Maturity (CMM, OPM3...) • Employees' satisfaction • Customers' satisfaction • Achievement of Objectives • Continuous Improvement
Individual Benefits: • Enhanced Competencies • Develop Management Skills • Foster New Opportunities • Value PM Experiences	• Project Management Career Path • Certifications • PM Trainings • PM Culture	• Employee satisfaction • Retention rate • Customer Satisfaction • Project Success Rate

Figure 9.1 Benefits, Principles, and Measurements of an Integrated Portfolio Management Framework

your own organization, here is your business case, and it's not only about the savings generated in terms of opportunity cost.

While looking at the table depicted in Figure 9.1, of course, you will notice that many of these elements are interdependent and overlapping. Indeed, they are! Portfolio management in the governance framework of an organization is the highest level of the pyramid (remember Figure 4.2). It's the sum of the proper implementation of all subsequent levels of operations, projects, and programs. It relies on the data and information gathered from the lower levels and needs these levels to be translated into reality.

9.4 Portfolio Management Implementation, Success Factors, and Prerequisites

One of the key success factors you will rely on when implementing portfolio management at a certain level of maturity is the already established maturity of the basic and fundamental elements of project and program management, and, of course, of operational steering. The implementation of these principles starts then from the bottom. Within the operations, basic information about resources have to be available. Of course, the resource demand planning exercise will help you in accessing and consolidating these data, but they must exist somewhere in the systems. Within the project layer, I often make sure that the basic rules of risk management are in place and that the performance of the projects can be effectively measured. Measuring performance implies being able to put in place a scope statement along with a work breakdown structure, and, if possible (that would be the icing on the cake), to spread the use of earned value management principles throughout the projects.

At the program level, we have to ensure that the concept of benefits and their management and monitoring is an accepted principle.

Implementing portfolio management is then a change. It's a change that starts from the lowest level of the organization and from the top level at the same time, and is aimed at connecting the dots. It requires time, communication, stakeholder management, competency development, and preaching. In other words, it requires the establishment of a dedicated program. You can look at the benefits detailed in Figure 9.1 to shape the benefit map of such a program, but these benefits will, of course, be specific to each organization and to the context in which it evolves. Not to mention that such a program has to be considered as part of the organization's portfolio, of course, thus triggering the loop it's a part of.

It's the aim of organizations, and it's the nature of people, to change. If we don't change, we die. The entire principle of evolution is based on that very statement. It's a matter of survival. And it's no different between organic beings or conceptual gatherings of individuals. Portfolio management is about harvesting the benefits and opportunities of the changes we face and not feeling threatened or scared by them. It's a means to sustain our development and move forward.

* * * * *

This last chapter has been completed on April 22, 2018, on board a plane, somewhere above the Atlantic, between Lisbon and New York.

Lexicon

This Lexicon contains the terms and concepts that I have taken the liberty to reformulate and, sometimes, to define or redefine. It's therefore not an exhaustive list of terms or concepts—just the ones that represent the key concepts presented throughout this book.

Organization: Any formally or informally defined cluster of co-organized, structured, and coordinated individuals working toward the achievement of a common objective—quantitative and/or qualitative. It can be temporary or not.

Program: A collection of harmonized and coordinated change actions (projects and activities) aimed at delivering benefits and capabilities not obtainable from individually framed initiatives.

Portfolio Management: The organizational governance layer that supports the realization of the strategy of that organization by optimizing the effective and efficient allocation of means and resources to maximize the generation of performance.

Organizational Portfolio: All the activities, present and potential, which imply the consumption of organizational resources and aim at contributing to the overall performance of an organization. That covers operations, projects, and programs.

Entropy: The tendency of any system to change its configuration from stability to chaos.

Uncertainty: A lack of quantitative information. Everything related to cost, time, effort, resources, etc. The less predictable are your quantities, the less accurate are your estimates; the more risks you have identified on a certain portfolio component, the more uncertain is that component.

Ambiguity: A lack of qualitative information. Everything related to the very definition of the final outcome, result, or deliverable of your portfolio component, or even about the definition of the process leading to produce that outcome.

Complexity: Uncertainty × Ambiguity. This represents the level of multiplication of unknown parameters and dimensions we will have to consider and manage when handling a specific component of our portfolios and/or the portfolios themselves.

Functional Organizational Structure: A vertical organizational structure based on the segments of activities and business operations conducted by the organization. Mainly adopted by companies generating most of their revenue through the product of their daily operations directly offered to customers who purchase them.

Projectized Organizational Structure: A vertical organizational structure based on the different projects conducted by the organization. Mainly adopted by organizations making all of their business by delivering project results to their clients.

Matrix Organizational Structure: A mixed organization, having vertical and transversal communication, decision, and authority channels. The fundamental architecture is the one of a Functional Organization, but the silos have been made permeable.

A **Weak Matrix** will be found in organizations where projects are undertaken to support the operations that remain the main source of business and profit.

A **Strong Matrix** will be established by organizations in which the main business focus is put on projects, but where the functions are used as resource providers supporting the projects.

Organizational Inertia: A lag in the implementation of strategic decisions in the organization. Expressed in terms of duration (*x* months, *x* weeks, etc.).

Organizational Agility: The ability of the organization to pursue its strategic vision and to realize it, while being able to anticipate the evolution of its business environment and adapt its strategic roadmap and related governance to this evolution.

Titanic Syndrome: The tendency for an organization to see threats and changes coming into its business environment, but at a time when it's already too late to react due to the level of organizational inertia, which inevitably pushes the organization to collide with that threat.

The main symptom of the Titanic syndrome is the panic mode into which the organization places itself, thereby wasting a tremendous amount of effort and resources in implementing desperate action plans that are supposed to be corrective but are, in fact, totally ineffective and inefficient.

Responsibility: Accepting ownership of specific activities within the defined roles and taking the initiative to deliver the agreed goals, objectives, and requirements of those activities. (Definition established in collaboration with Dr. Te Wu.)

Accountability: The ability to make empowered decisions within a defined perimeter of authority—delimited by the budget perimeter of the project, program, or portfolio—and to assume the consequences of these decisions.

There is an obvious overlap between these two components of commitment (accountability and responsibility). When assigned the responsibility for a certain activity, you also bear a certain accountability regarding the outcome of that activity. (Definition established in collaboration with Dr. Te Wu.)

Enterprise Management Office (EMO): A governance entity, formally established or diluted within various organizational entities assuming all or some of its roles, which is dedicated to supporting the consolidation of organizational data and providing strategic alignment as well as responsibility for propagating the components of that strategy throughout the organization.

Strategic Horizon: Time length within which the organization defines and realizes its strategy. The strategic horizon provides a temporal framework within which all active initiatives should be implemented, and their benefits earned and measured. It's calculated by taking into account the level of inertia of the organization and the evolution rate it faces in its business environment.

Business Environment Evolution Rate: The dynamic of your business environment at which pace your surrounding business environment changes. These changes can come from competitors, disturbers, or regulators, or the organization itself.

Strategic Vision: The expression of the desired configuration of the organization's market or environment in a defined future.

Organization's Strategy: The set of programs, portfolios, projects, and operations undertaken to realize the strategic vision of the organization.

Organization's Governance: The set of processes, tools, methods, and controls applied to conduct the realization of the organization's strategy.

Portfolio Manager: The individual whose accountability and responsibility consists of defining the portfolio strategy, translating it into tangible outcomes and performance objectives, and allocating organizational resources among a set of prioritized components (projects, programs, and operations) depending on their contribution to the portfolio's strategy and the expected performance, in an effort to maximize that performance by optimizing the use of organizational resources.

Stakeholder: Any individual, or group of individuals, directly or indirectly having an impact, or being impacted by, your initiatives, their processes, or their results.

Business Drivers: The strategic objectives at the enterprise level that represent the means and capabilities necessary for the organization to realize its strategic vision. Each initiative, at any level, undertaken within the organization should have, and will be measured against, a certain level of contribution to the achievement of one or more of these business drivers. Business drivers are strategic critical success factors (CSFs).

Critical Success Factor (CSF): A function having been identified as indispensable to the whole creation of value. CSFs represent the qualitative value creation indicators.

Portfolio Intrinsic Risks: Risks triggered by the execution of the portfolio components themselves. Also designated as portfolio internal risks.

Portfolio Extrinsic Risks: Risks mainly identified directly at the portfolio level, related to the business environment of the organization.

Risk Capacity: The financial ability of the organization to absorb a certain level of impact of risks. Risk capacity is represented by the sum of the reserve for incidentals and the risk–response budget.

Risk Appetite: The risk appetite represents the willingness of the organization (or in fact its executives and decision makers) to take a certain level of risk and expose the organization to a certain level of liability. The risk appetite is usually expected to be lower than the risk capacity.

Risk Tolerance: The risk tolerance is the difference between the risk appetite and the risk capacity. It determines the thresholds used to define the overall risk management strategies of the portfolio and its specific components.

References

Antoine de Saint-Exupery Quotes. (n.d.). BrainyQuote.com. Retrieved July 9, 2018, from https://www.brainyquote.com/quotes/antoine_de_saintexupery_103610

AXELOS. (2011a). *Management of Portfolios (MoP®)*. London, UK: The Stationery Office (TSO).

AXELOS. (2011b). *Managing Successful Programmes (MSP®)*—2011 Edition. London, UK: The Stationery Office (TSO).

AXELOS. (2017). *Managing Successful Projects with PRINCE2®*—2017 Edition. London, UK: The Stationery Office (TSO).

Bucero, A., & Englund, R. L. (2006). *Project Sponsorship*. San Francisco, CA, USA: Jossey-Bass.

Collective. (2001). Agile Manifesto. Retrieved December 1, 2017, from http://www.agilemanifesto.org

Edelman. (2016). Edelman Trust Barometer. Retrieved from https://www.edelman.com/trust-barometer

Gallup. (2013). State of the Global Workplace—2013 Report. Gallup.

Hobbs, B., & Aubry, M. (2010). *The Project Management Office (PMO): A Quest for Understanding*. Newton Square, PA, USA: Project Management Institute.

Johnson, G., & Scholes, K. (1997). *Exploring Corporate Strategy* (8th ed.). Upper Saddle River, NJ, USA: Prentice Hall.

Juran, J. M. (1964). *Managerial Breakthrough: A New Concept of the Manager's Job*. New York, NY, USA: McGraw Hill.

Lazar, O. (2010). The Project Driven Strategic Chain. PMI® Global Congress. Washington, DC, USA: Project Management Institute.

Lazar, O. (2011). Ensure PMO's Sustainability: Make It Temporary! PMI® Global Congress 2011—North America. Dallas, TX, USA: Project Management Institute.

Lazar, O. (2012). The Shift in the Project Management Profession: From Complication to Complexity. PMI® Global Congress EMEA 2012—Marseille, France. Newton Square, PA, USA: Project Management Institute.

Lazar, O. (2015a). "Aligning the Organization through Portfolio Management." Retrieved from https://learning.pmi.org/in-person-course/747/51/aligning-the-organization-through-portfolio-management/

Lazar, O. (2015b). Stop Reacting, It's Too Late! Start to Anticipate!: Use a Proper Budget Structure as an Early Warning System Integrating Cost, Scope and Risks. PMI® Global Congress EMEA 2015—London, England. Newton Square, PA, USA: Project Management Institute.

Lazar, O. (2015c). The Bricks for Building Your Portfolio: Risk, Benefits and Value. Portfolio Experience Conference. Warsaw, Poland.

Lazar, O. (2016). When Change Is Not a Change Anymore: Organizational Evolution and Improvement through Stability. PMI® Global Congress 2016—EMEA, Barcelona, Spain. Newton Square, PA, USA: Project Management Institute.

Lipke, W. H. (2012). *Earned Schedule*. lulu.com.

Merriam-Webster. (n.d.). Retrieved December 2, 2017, from https://www.merriam-webster.com/dictionary/inertia

Pfeiffer, J., Goodstein, L., & Nolan, T. (1986). *Applied Strategic Planning: A How to Do It Guide*. San Diego, CA, USA: University Associates.

PMI. (2017a). *A Guide to the Project Management Body of Knowledge* (*PMBOK® Guide*)—6th Edition. Newton Square, PA, USA: Project Management Institute.

PMI. (2017b). *The Standard for Portfolio Management*—4th Edition. Newton Square, PA, USA: Project Management Institute.

PMI. (2017c). *The Standard for Program Management*—4th Edition. Newton Square, PA, USA: Project Management Institute.

Porter, M. (2014). *Aligning the Strategy and Project Management*. PMO Symposium 2014. Phoenix, AZ, USA: Project Management Institute.

Sinek, S. (2009). *Start with Why: How Great Leaders Inspire Everyone to Take Action*. New York, NY, USA: Portfolio.

Thiry, M. (2004). "Program Management: A Strategic Decision Management Process." In P. W. G. Morris & J. K. Pinto (Eds.), *The Wiley Guide to Project Management* (Chap. 12). New York, NY, USA: John Wiley and Sons.

Thiry, M. (2013). *A Framework for Value Management Practice*—2nd Edition. Newton Square, PA, USA: Project Management Institute.

Thiry, M. (2016). *Program Management* (2nd ed.). Abington, Oxon, UK: Routledge/Taylor & Francis Group.

Tobak, S. (2009, March 25). Does Jack Welch Think Shareholder Value Is a Dumb Idea. Moneywatch. Retrieved from https://www.cbsnews.com/news/does-jack-welch-think-shareholder-value-is-a-dumb-idea/

Tuckman, B. W. (1965). "Developmental Sequence in Small Groups." *Psychological Bulletin, 63*(6), 384–399. Retrieved from http://dx.doi.org/10.1037/h0022100

Wikipedia. (2018, February 11). Temperature. Wikipedia.org. Retrieved from https://en.wikipedia.org/wiki/Temperature

Index

B

C

F

G

H

I

J

K

L

M

N

O

P

Q

R

S

T